LIST OF THE MISCELLANEOUS PUBLICATIONS
OF THE MUSEUM OF ZOOLOGY, UNIVERSITY OF MICHIGAN

Address inquiries to the Director of the Museum of Zoology, Ann Arbor, Michigan

Bound in Paper

No. 1. Directions for Collecting and Preserving Specimens of Dragonflies for Museum Purposes. By E. B. Williamson. (1916) Pp. 15, 3 figures $0.25
No. 2. An Annotated List of the Odonata of Indiana. By E. B. Williamson. (1917) Pp. 12, 1 map $0.25
No. 3. A Collecting Trip to Colombia, South America. By E. B. Williamson. (1918) Pp. 24 (Out of print)
No. 4. Contributions to the Botany of Michigan. By C. K. Dodge. (1918) Pp. 14 . . . . . . $0.25
No. 5. Contributions to the Botany of Michigan, II. By C. K. Dodge. (1918) Pp. 44, 1 map $0.45
No. 6. A Synopsis of the Classification of the Fresh-water Mollusca of North America, North of Mexico, and a Catalogue of the More Recently Described Species, with Notes. By Bryant Walker. (1918) Pp. 213, 1 plate, 233 figures $5.00
No. 7. The Anculosae of the Alabama River Drainage. By Calvin Goodrich. (1922) Pp. 57, 3 plates $0.75
No. 8. The Amphibians and Reptiles of the Sierra Nevada de Santa Marta, Colombia. By Alexander G. Ruthven. (1922) Pp. 69, 13 plates, 2 figures, 1 map $1.00
No. 9. Notes on American Species of Triacanthagyna and Gynacantha. By E. B. Williamson. (1923) Pp. 67, 7 plates $0.75
No. 10. A Preliminary Survey of the Bird Life of North Dakota. By Norman A. Wood. (1923) Pp. 85, 6 plates, 1 map $1.00
No. 11. Notes on the Genus Erythemis with a Description of a New Species (Odonata). By E. B. Williamson.
The Phylogeny and the Distribution of the Genus Erythemis (Odonata). By Clarence H. Kennedy. (1923) Pp. 21, 1 plate $1.00
No. 12. The Genus Gyrotoma. By Calvin Goodrich. (1924) Pp. 29, 2 plates $0.50
No. 13. Studies of the Fishes of the Order Cyprinodontes. By Carl L. Hubbs. (1924) Pp. 23, 4 plates $0.75
No. 14. The Genus Perilestes (Odonata). By E. B. Williamson and J. H. Williamson. (1924) Pp. 36, 1 plate $1.00
No. 15. A Check-List of the Fishes of the Great Lakes and Tributary Waters, with Nomenclatorial Notes and Analytical Keys. By Carl L. Hubbs. (1926) Pp. 77, 4 plates $1.50
No. 16. Studies of the Fishes of the Order Cyprinodontes. VI. By Carl L. Hubbs. (1926) Pp. 79, 4 plates $1.00
No. 17. The Structure and Growth of the Scales of Fishes in Relation to the Interpretation of Their Life-History, with Special Reference to the Sunfish Eupomotis gibbosus. By Charles W. Creaser. (1926) Pp. 80, 1 plate, 12 figures $2.50
No. 18. The Terrestrial Shell-bearing Mollusca of Alabama. By Bryant Walker. (1928) Pp. 180, 278 figures $1.50
No. 19. The Life History of the Toucan Ramphastos brevicarinatus. By Josselyn Van Tyne. (1929) Pp. 43, 8 plates, 1 map $0.75
No. 20. Materials for a Revision of the Catostomid Fishes of Eastern North America. By Carl L. Hubbs. (1930) Pp. 47, 1 plate $0.75
No. 21. A Revision of the Libelluline Genus Perithemis (Odonata). By F. Ris. (1930) Pp. 50, 9 plates $0.75
No. 22. The Genus Oligoclada (Odonata). By Donald Borror. (1931) Pp. 42, 7 plates $0.50
No. 23. A Revision of the Puer Group of the North American Genus Melanoplus, with Remarks on the Taxonomic Value of the Concealed Male Genitalia in the Cyrtacanthacrinae (Orthoptera, Acrididae). By Theodore H. Hubbell. (1932) Pp. 64, 3 plates, 1 figure, 1 map $0.75
No. 24. A Comparative Life History Study of the Mice of the Genus Peromyscus. By Arthur Svihla. (1932) Pp. 39 . . . . . . $0.50
No. 25. The Moose of Isle Royale. By Adolph Murie. (1934) Pp. 44, 7 plates $0.70
No. 26. Mammals from Guatemala and British Honduras. By Adolph Murie. (1935) Pp. 30, 1 plate, 1 map $0.35
No. 27. The Birds of Northern Petén, Guatemala. By Josselyn Van Tyne. (1935) Pp. 46, 2 plates, 1 map $0.45
No. 28. Fresh-Water Fishes Collected in British Honduras and Guatemala. By Carl L. Hubbs. (1935) Pp. 22, 4 plates, 1 map $0.25
No. 29. A Contribution to a Knowledge of the Herpetology of a Portion of the Savanna Region of Central Petén, Guatemala. By L. C. Stuart. (1935) Pp. 56, 4 plates, 1 figure, 1 map $0.50
No. 30. The Darters of the Genera Hololepis and Villora. By Carl L. Hubbs and Mott Dwight Cannon. (1935) Pp. 93, 3 plates, 1 figure $0.50
No. 31. Goniobasis of the Coosa River, Alabama. By Calvin Goodrich. (1936) Pp. 60, 1 plate, 1 figure $0.35
No. 32. Following Fox Trails. By Adolph Murie. (1936) Pp. 45, 6 plates, 6 figures $1.00
No. 33. The Discovery of the Nest of the Colima Warbler (Vermivora crissalis). By Josselyn Van Tyne. (1936) Pp. 11, colored frontis., 3 plates, 1 map $0.35

(CONTINUED ON LAST PAGES)

THE publications of the Museum of Zoology, University of Michigan, consist of two series—the Occasional Papers and the Miscellaneous Publications. Both series were founded by Dr. Bryant Walker, Mr. Bradshaw H. Swales, and Dr. W. W. Newcomb.

The Occasional Papers, publication of which was begun in 1913, serve as a medium for original papers based principally upon the collections of the Museum. The papers are issued separately to libraries and specialists, and when a sufficient number of pages has been printed to make a volume, a title page, table of contents, and index are supplied to libraries and individuals on the mailing list for the entire series.

The Miscellaneous Publications, which include papers on field and museum techniques, monographic studies, and other contributions not within the scope of the Occasional Papers, are published separately, and as it is not intended they will be grouped into volumes, each number has a title page and, when necessary, a table of contents.

MISCELLANEOUS PUBLICATIONS
MUSEUM OF ZOOLOGY, UNIVERSITY OF MICHIGAN, NO. 101

A Biogeography of Reptiles and Amphibians in the Gomez Farias Region, Tamaulipas, Mexico

BY

PAUL S. MARTIN

ANN ARBOR
MUSEUM OF ZOOLOGY, UNIVERSITY OF MICHIGAN
APRIL 15, 1958

Printed and bound by CPI Group (UK) Ltd, Croydon, CR0 4YY

Paperback ISBN : 978-0-472-75161-7

CONTENTS

ILLUSTRATIONS

PLATES

(Plates I-VII follow page 102)

FIGURES IN THE TEXT

MAPS

ACKNOWLEDGMENTS

IN 1948 I accompanied E. P. Edwards and R. P. Hurd on a four-month collecting trip through the Mexican states of Michoacán, Durango, and Tamaulipas. The discovery of undisturbed Cloud Forest in northeastern Mexico encouraged me to develop a regional study of this area.

Since then so many individuals have contributed information, specimens, or other assistance that I suffer an embarrassment of riches in their acknowledgment. Field work in the Gómez Farías region was made pleasant and profitable through the inspiring co-operation of William Francis (Frank) Harrison and Everts Storms. The extent to which B. E. Harrell, C. F. Walker, and G. M. Sutton have shared their knowledge of this area deserves special notice. Finally, I have enjoyed the sustained encouragement and resourceful companionship of my wife, Marian W. Martin.

A BIOGEOGRAPHY OF REPTILES AND AMPHIBIANS IN THE GOMEZ FARIAS REGION, TAMAULIPAS, MEXICO*

INTRODUCTION

TWO centuries after Hernando Cortés subdued the Aztec Empire, that part of northeastern Mexico between the Río Grande and the Río Tamesí still resisted conquest. From strongholds in the northern part of the Sierra Madre Oriental known as the Sierra Gorda, hostile Indian tribes waged incessant guerrilla warfare. Such Spanish settlements as Pánuco, Querétaro, Matehuala, and Saltillo invited attack. From Tampico to Texas there was no point on the frontier that did not witness the "ravages of the barbarian" (Hill, 1926:52). Finally, in January of 1747 José de Escandón led 765 troops in a successful campaign of pacification. Within two years the present towns of Llera, Ocampo, Xicotencatl, Ciudad Victoria, and at least ten others had been founded or rebuilt.

With the advent of Spanish colonization the natural vegetation and native fauna, previously subjected to Indian agriculture, burning, and hunting, experienced a new level of cultural disturbance. By 1757, ten years after Escandón's entry, Tienda de Cuervo censused the young province of Nuevo Santander and found it had grown to 8,000 colonists, approximately 80,000 cattle, and 300,000 sheep (Hill, 1926:9). It is not the nature of this cultural shift but its initial two hundred years delay that is unusual in the history of colonial Mexico.

In similar fashion northeastern Mexico was neglected by early collectors and scientists. During the last hundred years scattered collections of plants and various animals were assembled by the Mexican Boundary Commission, L. Berlandier, W. W. Brown, E. A. Goldman, F. Armstrong, F. W. Pennell, and H. A. Pilsbry. Not until the construction of the Laredo-Mexico City Highway in the early 1930's were any systematic studies completed. By comparison with the much-traveled, much-collected part of central Mexico between the cities of Puebla and Veracruz, the biological description of border tropical habitats in northeastern Mexico is quite recent.

In analyzing ecological distributions of reptiles and amphibians in southern Tamaulipas I have confined observations to a small, if topographically complex, section of the Sierra Madre Oriental. This method enables a more careful definition of zonal distribution than would be possible had the same amount of field work been expended in a larger geographical unit. The area chosen lies in southwestern Tamaulipas

*A revised version of a dissertation submitted in partial fulfillment of the requirements for the degree of Doctor of Philosophy at the University of Michigan, 1955.

Accepted for publication, September 17, 1956.

immediately south of the Tropic of Cancer between 22° 48′ and 23° 30′ N. lat. and between 99° and 99° 30′ W. long. The Municipio of Gómez Farías lies entirely within these parallels, and I shall designate the quadrangle thus enclosed the Gómez Farías region (Map 1).

Three important tropical plant formations, Tropical Deciduous Forest, Tropical Evergreen Forest, and Cloud Forest, are unknown north of this region. Thus, the region provides opportunity to study these formations and their faunas under limiting environmental conditions. Hooper (1953) considered the eastern part of San Luis Potosí and southern Tamaulipas in the following light: "From a zoogeographical standpoint it is perhaps more important as a region of transition, where tropical faunas, floras, and climates impinge on and give way to temperate environments."

The part of northeastern Mexico through which biogeographers since the time of Sclater and Wallace have drawn the line separating Temperate (Holarctic or Nearctic) from Tropical (Neotropical) regions is southern Tamaulipas. In this regard the observations of Salvin and Godman (1889) are of interest: "From this it will be seen that the line of demarcation between the two regions [Nearctic and Neotropical], so far as the birds are concerned, is capable of being defined with some precision, and will be found to coincide with the northern limits of the forests. Those on the eastern side leave the coast a little north of Tampico, and continue in a narrow belt along the eastern flank of the mountains in a nearly northern direction almost to Monterrey."

The term forests in this case is used in a very broad sense and presumably includes Tropical Deciduous Forest near Tampico and oak forest of the foothills near Monterrey.

In addition to its critical geographical position a second reason for selecting the Gómez Farías region was its relatively primeval state. Until 1950 much of the region, perhaps more than 50 per cent, was covered with natural forest, ostensibly climax or near climax. Most of the mountains between Chamal and Carabanchel (Map 1) were uninhabited. Possible interference by prehistoric man is difficult to evaluate. Historical records and archeological finds indicate extensive Indian occupation of the area including the interior valleys and the montane forests. Allegedly a mission, Mision de la Sierra de la Soledad de Igoya, was active in the mountains west of Gómez Farías in early post-Conquest time. Whatever the influence of this venture, human activity in the mountains in recent years was largely restricted to three or four small ranches and settlements. Until 1951 when intensive lumbering began, the total population of the Sierra Madre between Gómez Farías and La Joya de Salas north to Carabanchel did not exceed twenty families.

Since 1951 lumbering has destroyed or drastically modified much of the montane forests between 900 and 2400 m. Although lumber roads facilitated travel in parts of the mountains marked "inaccessible" on an earlier map (Heim, 1940), it is regrettable that a more intensive biotic survey could not be completed before such disturbance.

Despite agricultural activity in the lowlands and interior basins the intensity of human pressure in the Gómez Farías region is not comparable to that in the Huastecan district of eastern San Luis Potosí. It is still

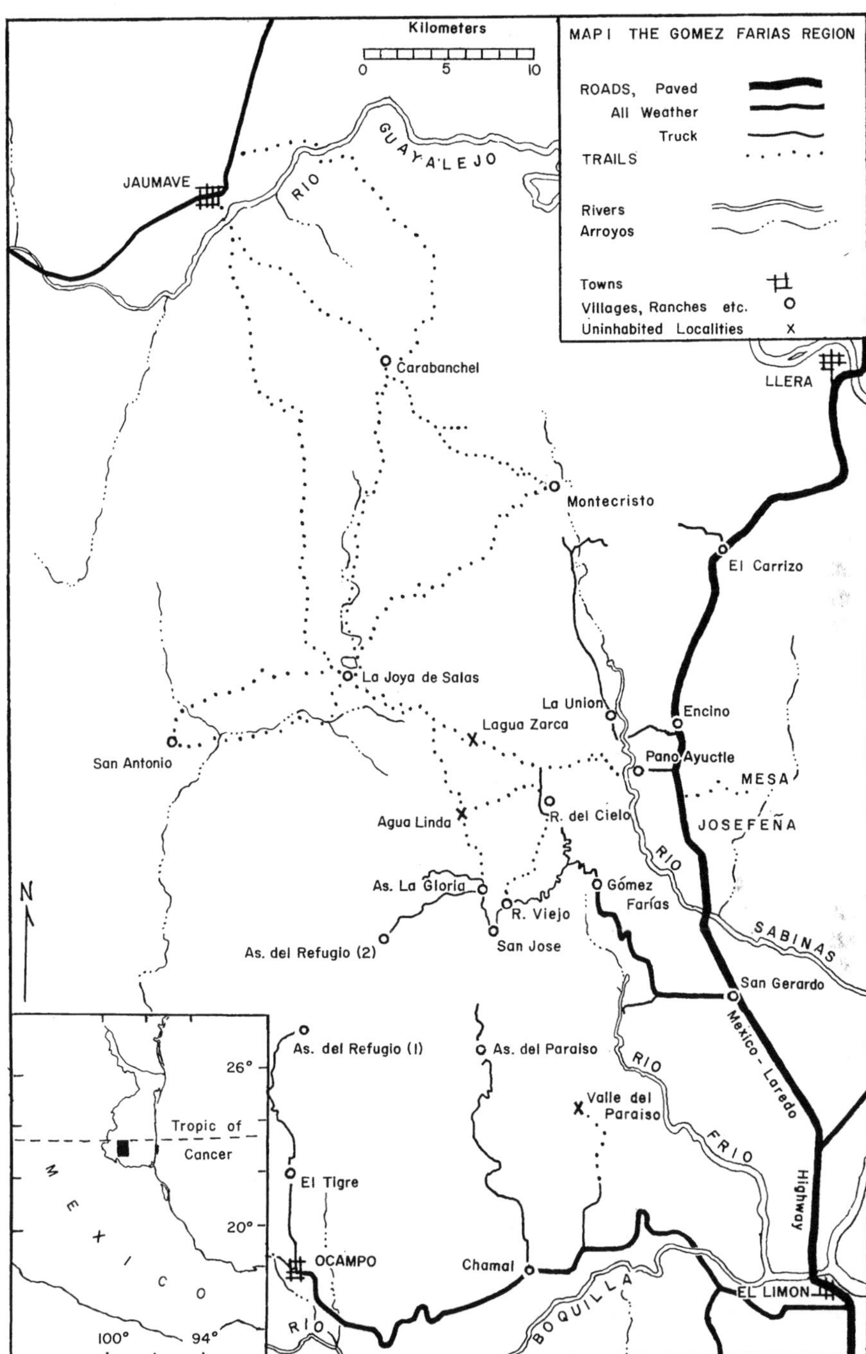

Map 1. The Gómez Farías region, Tamaulipas, 22° 48′ to 23° 30′ N. lat. and 99° to 99° 30′ W. long. Only localities visited and routes traveled are shown. Rancho is abbreviated by the letter R, Aserradero (sawmill) by As.

possible to reconstruct the major natural features of the Tamaulipan landscape.

A third advantage in selecting this region was the number of field collectors and investigators that have visited the area. Not only have they supplied a very large part of the total reptile and amphibian collections, but their observations and reactions to the region have aided me greatly in my own interpretations. The more important herpetological collections are listed in Table I; in addition, a few specimens have come from others incidental to their major interests. In this regard, W. F. Harrison has supplied many important records.

TABLE I

Source of the Larger Herpetological Collections from the Gómez Farías Region

Date	Localities	Collectors	No. of Specimens
1948			
Apr.5-May 16	Rancho del Cielo	E. P. Edwards	
June 11-18	Pano Ayuctle	R. P. Hurd	
		P. S. Martin	31
1949			
Feb.23-Apr.1	Pano Ayuctle	C. R. Robins	
May 18-30	Rancho del Cielo	W. B. Heed	
	La Joya de Salas	P. S. Martin	
	Chamal		528
1950			
Apr.1-June 7	Pano Ayuctle	R. M. Darnell	
	Rancho del Cielo	B. E. Harrell	*ca.* 100
July 27-Aug.5	Rancho Viejo	M. and P. S. Martin	
	Rancho del Cielo		77
Aug.14-Sept.4	Rancho del Cielo	W. B. Heed	
	Lagua Zarca	C. F. Walker	
	Pano Ayuctle		215
1951			
Mar.13-June 15	Pano Ayuctle and vicinity	R. M. Darnell	
	La Joya de Salas		*ca.* 75
June 17-Aug.3	Gómez Farías	W. Z. Lidicker	
	Pano Ayuctle	J. Mackiewicz	
	La Joya de Salas	M. and P. S. Martin	
	Rancho del Cielo		428
1952-53			
Dec.21-Jan.3	Pano Ayuctle	R. M. Darnell	
	La Union	E. A. Liner	
	Rancho del Cielo		212
1953			
Feb.4-Mar.23	All localities	P. S. Martin	
Apr.5-June 24	shown on Map 1	B. E. Harrell	
		C. F. Walker	828
Total 14 months between 1948-1953			*ca.* 2500

THE ANIMAL ENVIRONMENT

Geology, climate, and vegetation are environmental features of primary concern to the animal ecologist. To facilitate an understanding of animal habitats in the Gómez Farías region I shall discuss each of these in turn. Such information should clarify the environmental basis for certain distribution patterns both throughout eastern Mexico and, locally, in the Gómez Farías region. In addition it should be useful in comparing this with other peritropical areas.

Within the Gómez Farías region I have found vegetation the best environmental index. Perhaps it would be preferable to describe animal habitats in climatological, edaphic, and other physical terms. Unfortunately, standard meteorological data are not sufficient to define more than the major latitudinal climatic gradients in Mexico. The vertical gradients are very poorly known; nothing comparable to Brown's Philippine mountain study (1919) has been attempted. Allee (1926) and a few others have made a start, but microclimate is virtually an untouched field in the New World tropics. Other physical features, such as light, soils, evapotranspiration, heat transfer of the substrate, require much more refined measurement than was practicable in the present study.

Some system of classifying animal habitats is important. While scarcely a panacea, an understanding of the relationship between vegetation and animal habitats is useful. I have selected the Plant Formation (Schimper, 1903), or Vegetation Zone (Leopold, 1950), as the most effective measure of environmental similarity. Plant formations are assumed to respond primarily to climatic controls (Dansereau, 1952:325). In addition, I assume that within a single plant formation similar paths of energy transfer, microclimates, and shelter types are found. Certainly the terrestrial vertebrates and other animals can select various microenvironments and thus avoid the environmental extremes experienced by the Plant Formation; the formation, however, limits the number and nature of these microenvironments. It is used, therefore, as an index of potential animal habitats, regardless of whether these are filled in any given area.

From a knowledge of formations or vegetation zones various biogeographical questions arise which might otherwise escape notice. Are Humid Pine-Oak Forests of eastern and western Mexico ecologically equivalent? If so how can the variety of Plethodontid salamanders in the former, four genera and over 25 species, and their apparent absence in the latter be explained? Why is the fauna of Tropical Deciduous Forest and Thorn Forest formations in the northern end of the Yucatán Peninsula richer in species derived from Tropical Rainforest genera than similar formations found in Tamaulipas and San Luis Potosí? Does the Cloud Forest fauna on either side of the Isthmus of Tehuantepec demonstrate the result of Pleistocene interconnections? Such questions require information on the nature and distribution of vegetation types and their associated faunas.

In addition to historical problems I use vegetation types to indicate zonal change. Although empirical, such a method may prove less arbitrary than the life zone system used by some zoogeographers (Dalquest,

1953; Goldman, 1951; Griscom, 1950). Recognition of Mexican life zones has been based largely on vertebrate indicators. This approach anticipates my object, e.g., to characterize zones in terms of vegetation first and then to define the degree of faunal fit. For example Cloud Forest (Subtropical or Humid Upper Tropical Zone) in this area lacks absolute animal indicators among the vertebrates, despite its distinctive floristic and vegetational nature (Martin, 1955b).

Thus, in the assumption that information on local animal distribution is most useful when related to vegetation zones, I have sought to record such data for the reptiles and amphibians of the Gómez Farías region. Clearly the present survey is incomplete; presumably other studies along altitudinal gradients elsewhere in northern Middle America will help define which species "fit" a particular environmental pattern either locally or regionally.

Geology of the Gómez Farías Region

Those aspects of geology of interest to the ecologist include: (1) the effect of physiography on climate; (2) the effect of substrate on animal and plant habitats; and (3) the role of historical geology. Because of its proximity to the rich Mexican Northern Fields, the geology of eastern Mexico pertinent to petroleum exploration is relatively well known. In the following summary I have drawn largely upon Muir (1936), Kellum (1930), and especially Heim (1940), all of whom treated the Gómez Farías region in their accounts of northeastern Mexico.

Physiography and Climate

Heim described the Sierra Madre Front between Ciudad Victoria and Llera as a series of anticlines that rise gradually to the west. "They are secondary folds on the easterly limb of the main anticline and dye [*sic*] out like waves toward the great synclinal plain of Ciudad Victoria" (Heim, 1940, pp. 335-336).

The main front of the Sierra Madre continues south of Victoria, rising to a broad plateau of 2100 m. at Carabanchel, with isolated peaks near Agua Linda exceeding this. One, called Mount of Oaks, is said to reach 8000 feet (2400 m.) according to Everts Storms. Between Victoria and Gómez Farías this broad fold is broken by a narrow gap cut by the Río Guayalejo, which debouches from the Jaumave Valley through a gorge onto the coastal plain west of Llera. Heim terms the part of the Sierra Madre Oriental south of the Guayalejo gap the Carabanchel Anticline. Kellum (1930:89) agrees that the secondary folds "which rise *en echelon* north of Buena Vista ranch" (near Encino on Map 1) are definitely anticlinal (p. 89) but he considers that the front of the escarpment has been formed as an overthrust of El Abra limestone overriding Tamaulipas limestone. This elevated, precipitous segment of the escarpment, which Heim calls the Carabanchel Anticline and Kellum considers a part of the overthrust Abra-Tanchipa front, is known on old property maps as the "Sierra de

Guatemala" (Sharp *et al.*, 1950; Hernández *et al.*, 1951; Harrell, 1951). I follow their usage, here restricting the term to that part of the Abra-Tanchipa front of the Sierra Madre Oriental between the Guayalejo gap and the Chamal valley.

South of the vicinity of Llera two important structural changes occur. The first is the plunging and disappearance of the minor front anticlines. West of Pano Ayuctle the precipitous frontal ridge of the Sierra de Guatemala rises abruptly from the lowlands, unobstructed by foothills. At Gómez Farías only one isolated anticline of about 700 m. elevation stands in front of the wall-like escarpment. "The sixth range westward from the front of the Sierra Madre, southwest of Victoria, becomes the front range (Sierra del Abra) toward the south" (Muir, p. 159).

The second important change involves the deformation of this front range south of San José, which Heim illustrated (Fig. 7). "In the Chamal region, most interesting changes in structure occur. The wide syncline at this village lies exactly on the southern projection of the Carabanchel Anticline. The wide Tamabra-built mountain gradually descends to the south, the anticlinal crest being transformed into a syncline and the flanks into lateral anticlines" (Heim, p. 338).

From Chamal south the Sierra Madre front continues to the Río Tampo in San Luis Potosí as a low, narrow, comparatively simple fold, various sections of which are known as the Sierra del Abra, Sierra Cucharas, and Sierra Tanchipa (Kellum, 1930). Behind this low front one or two other low ridges also precede the Plateau Escarpment. Between the Gómez Farías region and the Xilitla region of San Luis Potosí the passes into the Plateau are quite low, not exceeding 1430 m. between Ocampo and Tula, and no higher between Antiguo Morelos and Ciudad Maiz in San Luis Potosí. The Plateau Escarpment may not exceed 1800 m. at any point between the Sierra de Guatemala of the Gómez Farías region and Cerro Conejo (2650 m.) west of Xilitla.

From the two tectonic changes described above one may conclude that the elevated, broad Sierra de Guatemala provides a formidable barrier to rain-bearing easterlies. The orographic effect produced by this part of the Sierra Madre front appears to be unequaled elsewhere in northeastern Mexico. No similar barrier is known between Xilitla and Ciudad Victoria, a distance of 250 km.

Substrate and Its Biological Effect

The geological formations described by Heim and Kellum include three of Cretaceous age, the Tamabra limestones, the San Felipe limestones, and the Mendez shale. Cenozoic igneous deposits and lowland alluvium complete the list. Each of these has its own effect on vegetation, although the difference between the San Felipe and Tamabra may be very slight.

The El Abra limestone, a facies of the Tamabra, which makes up the bulk of the Sierra de Guatemala, presents entirely different erosional surfaces on its eastern (steep) and western (less precipitous) sides. USAAF Trimet photo (2-4008) (1-L109) (2B) shows the minute details of gully incision and valley drainage via surface runoff on the west side of the Sierra de Guatemala. By contrast, the east side as seen from the lowlands or along mountain trails exhibits virtually no trace of surface drainage. The entire front between Gómez Farías and Montecristo is severely folded. A karst topography in late youth is developed on this surface. A variety of karst forms including caves, sinkholes of various diameters and depths, pinnacles, uvalas, and haystacks characterize the surface. So efficient is the drainage that in the dry season very few natural springs and only two short permanent streams (Agua Linda and Ojo de Agua de los Indios) are known. Even during the rainy season surface drainage is temporary. After torrential storms in early August at Rancho Viejo I have seen clear surface streams of considerable size develop, then disappear within two days. Dan Cameron of Chamal recalls that when he lived in the valley (apparently an uvala) of what is now Ejido Alta Cima, the bottom of this depression would fill to a depth of several meters with clear water after an unusually heavy storm.

Drainage on the east slope therefore is entirely subterranean to the foot of the Sierra where two large springs form the headwaters of the Río Sabinas and Río Frío, respectively (Map 1).

The absence of permanent surface water severely limits the pond- and stream-breeding Anura. Artificial water holes (tanques) for cattle and a few springs hold enough water to allow *Bufo*, *Rana*, *Smilisca*, and *Hyla* to breed in this area, but all of these are scarce, especially on the east slope of the mountains.

The wealth of caves, crevices, and sinks, however, affords a great variety of subterranean habitat for those salamanders and frogs that undergo direct development and are not dependent on surface water. The importance of caves as amphibian habitats at all seasons cannot be overemphasized.

The effect of lapies and spires will be discussed in the section on vegetation.

In the lowlands the areas covered by shale (Mendez) are usually edaphically drier than adjoining alluvial or limestone areas. Associated with these shale outcrops one may expect the lizard *Holbrookia texana* and various xeric shrubs such as yucca and organ pipe cactus. These grow in areas surrounded on better soils by Tropical Deciduous Forest.

The alluvial lowland soils are important in the development of "palm bottom," which is found almost exclusively over deep, black earth. Many such areas are now under cultivation, and the rest are rapidly being cleared.

Lava soils are found locally in the Gómez Farías region. Where well watered they are favorable to agriculture, and the small area of lava soil about Gómez Farías is sufficient to make this village one of the most productive in native tropical agriculture in southwestern Tamaulipas. The mesa tops including Mesa Josefeña are capped with basalt and are largely grass-covered. A lava stream followed the Río Boquilla gap from the Ocampo Valley through the Sierra de Chamal into the Chamal Valley and formed a dike in the palm bottom near the Cameron ranch. Heim (p. 335) suggested that this lava stream may have run in recent times.

Volcanic plugs that dot much of the coastal lowlands and reach their largest size in the spectacular Bernal de Horcacitas (1111 m.) east of Ciudad Mante do not occur in the Gómez Farías region.

Historical Geology

The Paleozoic and especially the Mesozoic history of most of northeastern Mexico has been described in detail by a variety of petroleum geologists (see especially Muir's thorough review and extensive bibliography). Unfortunately, post-Mesozoic events remain largely unknown or undescribed. Of particular interest to the biogeographer is the age of epeirogeny and the subsequent history of the Mexican Plateau.

Despite lack of confirmation from other authorities, the views of Schuchert (1935) concerning recent elevation of the Plateau continue to prevail. Thus, Sharp (1953), faced with the problem of origin of the Mexican temperate flora, commented: "*If Schuchert is right* [italics his], it seems clear that following the appearance of the Angiosperms, Mexico had little area of sufficient elevation continuing through geological time to support temperate vegetation until the Pliocene."

In assembling evidence for late Tertiary or Pleistocene migration of moss floras between eastern United States and humid montane forests in Mexico, Crum (1951) also cited Schuchert as the authority for relatively recent elevation, and hence recent development, of Mexican montane habitats.

With regard to Mexico, as in most of Middle America, Schuchert held that the main diastrophism began in late Pliocene: "The Mesa Central was eroded into the Cordilleran peneplane during Cenozoic time... Finally, in the late Pliocene and during Pleistocene came the very great epeirogenic elevation which produced present Mexico, elevating the land 3000-4000 feet in the north and 7000-8000 feet south of Mexico City (p. 133). ... Following this [Cordilleran] peneplanation, another uplift began, probably in late Pliocene time, was most active in the Pleistocene, and 'is still in progress' " (p. 124).

Other geologists are largely noncommittal with regard to uplift; however, they do place orogeny (folding) at an earlier time than the Pliocene. Muir (1936:140) considered the Tertiary history thus: "The beginning of orogenic movements in the Sierra Madre appears to post-date the Upper Midway [Lower Eocene]. The time of this orogeny is probably not older than early Wilcox time. ... As the Oligocene deposits show the effect of considerable movement, it seems likely that the later phases of the orogenic movements in the Sierra Madre lasted until about the beginning of Miocene."

Heim (1940) was less specific: "The main folding of the Front Ranges is post-Chicontepec or post-Paleocene. It terminated before the lava-flows of the unfolded mesas, whose present elevation is due to recent uplift."

Kellum (1937:35) reviewed the history of northeastern Mexico as follows: "Böse and Cavins consider the folding of the San Carlos Mountains and the Sierra de Tamaulipas to be older than that of the Sierra Madre Oriental and ranges west of it. The former were lifted above the sea in

Campanian time and were arched during Maestrichtian time [n.b. both Upper Cretaceous]. The Sierra Madre Oriental was lifted in the Maestrichtian, but the main folding occurred in the lower and middle Eocene."

More recently a vertebrate fossil deposit of late Eocene or Oligocene age in the Guanajuato red conglomerate "permits for the first time a correlation between the orogenic history as far west as Guanajuato and that along the Gulf coast east of there, indicating that the major compressive orogeny inland had been effected also by late Eocene time" (Fries, Hibbard, and Dunkle, 1955). The conglomerates are further described by Edwards (1955), who considers their red color as evidence of a humid temperate climate in the early Tertiary of this region.

Stirton's (1954) discovery of a late Miocene horse in sedimentary beds on the Isthmus of Tehuantepec indicates both an elevated mountain mass eroding to form sediments and savanna conditions in the area, presumably on the dry side of a range.

Biologically speaking, the most serious objection to a strict Schuchertian interpretation is that it requires extremely rapid evolution of the modern Plateau biota, essentially a temperate, semi-xeric to mesic assemblage. The Plateau is widely recognized as an evolutionary center of profound influence on the North American continent. A few outstanding examples of large genera with centers of differentiation in the Plateau include: *Thamnophis, Sceloporus, Crotalus* (reptiles); *Reithrodontomys, Neotoma, Thomomys* (mammals); *Aphelocoma, Aimophila, Pipilo* (birds); *Humboldtiana* (snails). In addition to the preceding autochthonous genera the Plateau and its escarpments harbor an impressive number of narrowly endemic, usually monotypic, genera such as the following: *Toluca, Conopsis, Adelophis* (reptiles); *Nelsonia, Romerolagus, Neotomodon* (mammals); *Xenospiza, Plagiospiza,* and *Ridgwayia* (birds). Finally, the great variety of pines (39 species of *Pinus* listed by Martinez, 1945) and oaks (112 species of *Quercus* listed by Standley), most of these confined to the uplands, strongly suggests an important, and enduring, secondary center of evolution for these genera in the Mexican highlands. It appears unlikely that these groups could have evolved from tropical or subtropical lowland progenitors in roughly one million years (length of the Pleistocene). That they could have evolved elsewhere and dispersed into this area either from temperate areas farther north or from another plateau to the south, without leaving a trace of their former origin in the form of fossil evidence or relic distribution, seems equally improbable. Thus we find both geological and biological grounds for assuming a longer history of uplift of the Central Plateau than assigned by Schuchert.

Summary

This brief discussion treats only a few aspects of the geology of the Gómez Farías region. It should facilitate an understanding of the climate, vegetation, and history of the area. The Sierra Madre front represented near Gómez Farías by the Sierra de Guatemala rises abruptly from the coastal plain to 2400 m., unobstructed by foothills and exposed on both south and east flanks. It forms a major topographic barrier to the easterly trade winds which produce a maximum of orographic precipitation on these slopes. The weathering of Cretaceous limestones that form the front is important in influencing development of vegetation and in providing various animal habitats. Karst topography is an outstanding feature of east slopes of the mountains. Soil is confined to shallow pockets, crevices, and the bottom of a few basins. In the lowlands both Cretaceous limestones and shales are present along with more recent alluvium and lava intrusions.

Various views on the age of the Sierra Madre Oriental and the Mexican Plateau are mentioned. I find biological and geological evidence for viewing the history of uplift of the Central Plateau as antedating the late Pliocene.

Climate

The following discussion is a brief synopsis of certain climatological features, emphasizing those assumed to be of importance to animals and plants.

Although not abundant, climatological data for northeastern Mexico are sufficient to describe certain relatively homogeneous areas such as the Gulf Coastal Plain and parts of the Mexican Plateau. On the other hand, they are quite inadequate to describe the great variety of local climatic gradients produced by the orographic effect of the Sierra Madre Oriental and the coastal plain ranges. Diversity of climate in these areas may be recognized in terms of vegetation, a method employed with notable success by Muller in Nuevo León (1937, 1939) and Coahuila (1947). Muller's forceful statement (1939:693) that "... existing climatological classifications are worse than worthless to an ecologist or geographer working in mountainous regions" applies generally to northeastern Mexico. The importance of mountain chains as biotic highways in this region makes this problem the more acute. The humid montane climates described by Muller in Nuevo León and Coahuila and represented with some additions in the Gómez Farías region remain virtually uncharted on existing climatic maps (Contreras, 1942; Tamayo, 1949; Vivo y Gomez, 1946). Problems of cartography as well as a lack of data may justify their omission.

Before discussing the local climatic types found in the Gómez Farías region, I shall consider the general pattern of sea level gradients encountered in eastern Mexico and southern Texas. Data for the latter are derived primarily from Contreras (1942) and from Mills and Hall (1949).

Climatic Gradients in Eastern Mexico

On the Gulf Coastal Plain between latitudes 18° and 29° N. a variety of tropical environments encounter limiting conditions. A formation series from Tropical Rainforest of southern Veracruz through various Evergreen and Semi-Evergreen Seasonal Forests, Savanna, Tropical Deciduous Forest, Thorn Forest, Thorn Scrub, and finally Coastal Prairie of southeast Texas is encompassed in this distance. Centuries of cultivation and other human interference have destroyed or disturbed the coastal plain forests, but the original sequence between Coatzacoalcos and Brownsville probably resembled the Seasonal Formation Series illustrated by Beard (1955:91, Fig. 2).

Temperature. — Along the Mexican coasts there is slight decrease in mean annual temperature from south to north (Ward and Brooks, 1938:J13). In Gulf coastal stations, those least affected by proximity to the Sierra Madre and those which generally have reliable records, a drop in mean annual temperature of 4.1° C (7.3° F) appears through the 11 degrees of latitude between Coatzacoalcos and Galveston (Fig. 1). The northward spread between means of the three warmest and three coldest months is of interest. At Coatzacoalcos the difference is 4.4°, increasing to 8.7° at Tampico, 12.2° at Brownsville, and 15.6° C at Galveston.

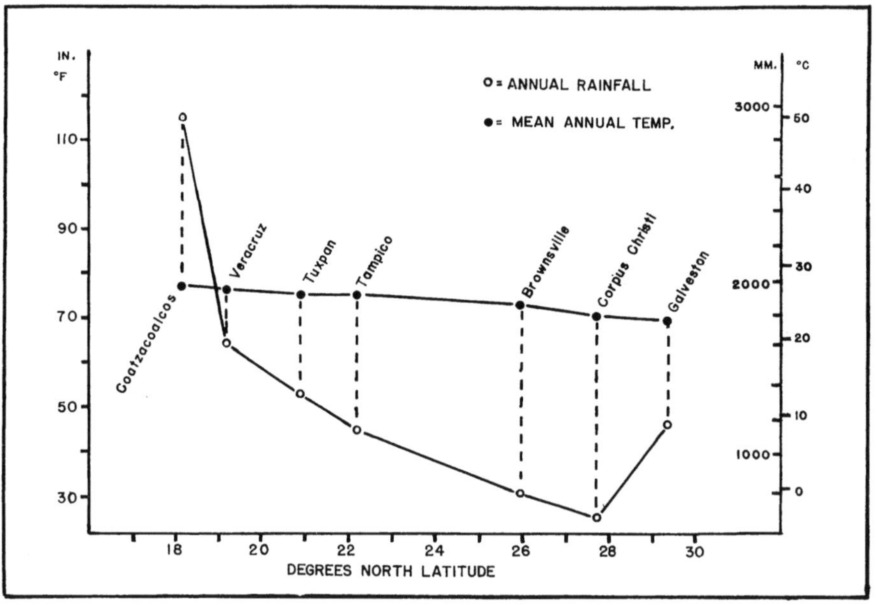

Fig. 1. Rainfall and temperature gradients along the Gulf of Mexico between 18° and 30° N. lat. Data from Mills and Hall (1949) and U.S.D.A. Summary of the Climatological Data for the U.S. by sections.

Along Atlantic coastal stations within the next 10° of latitude north of Galveston, roughly from Jacksonville to New York City, the fall in mean annual temperature is 10.0° C (18° F).

In the Köppen system the mean of the coldest month is weighted in defining climatic boundaries, rather than mean annual temperature. This factor would seem to be of considerable importance in controlling distribution of a tropical biota. A monthly mean below 18° C is considered the dividing point between temperate (C) and tropical (A) climates, and this line is drawn across the Gulf Coastal Plain just south of the Tropic of Cancer (Köppen in Ward and Brooks, 1938: J46). Such a definition emphasizes the continental influence of polar outbreaks ("Nortes" or Northers), which sweep down the Gulf Coast in winter. Typically the "Norte" brings cold, usually dry air, heavy winds, cloudy skies, and occasional squalls. Wind-blown soil may tint the sky a light brown or create dust haze. The "Norte" lasts one to several days before warm weather, clear skies, and humid air masses return.

Unusually severe outbursts, as occurred in early February, 1951, after a near record barometric high of 1065 mb. over northwestern Canada (Miller and Gould, 1951), produce killing frosts. On this occasion freezing temperatures of record-breaking duration occurred in southeast Texas (Farrell, 1951) with 88 consecutive hours of freezing weather in parts of the lower Río Grande Valley. At Rancho del Cielo on February first Frank Harrison recorded a minimum of -6° C and noted considerable damage to vegetation, especially to epiphytic plants (Hernandez *et al.*, 1951). Two years later the large tank bromeliads characteristic of this forest prior to 1951 (Pl. V) were still quite scarce although they remained fairly common in the Pine-Oak Forest above 1450 m.

In the lowlands damage was extensive. Although few trees were actually killed, the aftereffects of severe frost pruning were noted at Pano Ayuctle and Gómez Farías in July, 1951 (Pl. II, Fig. 2). Despite heavy summer rains new growth was so sparse that many areas of Tropical Deciduous Forest retained a dry season aspect with gray trunks and leafless branches. Two years later damage was still evident on many trees. Observations made in the summer of 1951 indicated frost pruning at least as far south as the Xilitla region of San Luis Potosí.

Goldman (1951:233) described the result of a freeze that killed trees in "Humid Lower Tropical Zone" at Metlaytoyuca, Veracruz (240 m. elevation) several years prior to his visit in 1898. Undoubtedly such extreme winter conditions are rare events, but they and the more frequent, less severe "Nortes" must play an important role in limiting the northward spread of the tropical biota in eastern Mexico.

Polar outbreaks are largely confined to the Atlantic Slope. North of the Isthmus of Tehuantepec the Sierra Madre Occidental and Sierre Madre del Sur shelter the Pacific Slope. Garbell (1947) presented both a theoretical and a regional consideration of this phenomenon. One of three requirements for a sudden polar outbreak reaching low latitudes is "a high mountain system to the west of the affected region" (Garbell, 1947:68).

Rainfall. — Richards (1952) considered the latitudinal limit of Rainforest to be under pluvial rather than thermal control. This factor definitely limits the extent of Rainforest in eastern Mexico. It also controls the sequence of other tropical vegetation types along the Atlantic lowlands, although an interaction with temperature assumes increasing significance northward.

Unlike the thermal gradient, relative change in mean annual precipitation between latitude 18° to 29° is rapid (Fig. 1). Rainfall is thus assumed to exert the primary control on northward spread of tropical forests. The mean annual precipitation falls from 2920 mm. at Coatzacoalcos to 640 mm. at Corpus Christi. Muller (1939:711) demonstrated northward decline in precipitation along a small part of this gradient in Nuevo León. A mean annual precipitation of 409 mm. is reported at Nuevo Laredo, which lies in an arid wedge of B type Köppen climates roughly equivalent in area to the lower Río Grande Valley. This is the driest part of the Gulf Coastal arc outside of the Yucatán peninsula. This dry wedge forms a major barrier between humid tropical and humid temperate climates on either side in southeastern Mexico and southeastern United States.

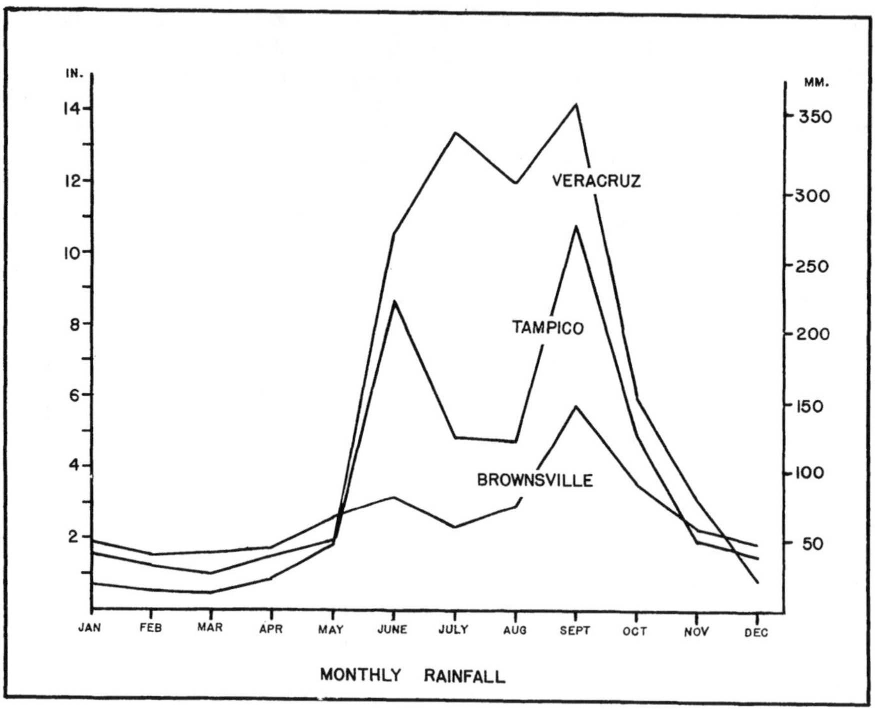

Fig. 2. Rainfall distribution at Veracruz, Tampico, and Brownsville. The summer rains diminish in intensity northward as the annual distribution changes from a seasonal tropical to a continental temperate regimen.

The decline in precipitation northward along the Gulf Coast is attributed to the diminishing effect of the Caribbean trades, and the gradual inland trend of the Sierra Madre Oriental and its lower elevations north of Monterrey. The gradient is not perfectly smooth and in Figure 1 it is idealized to some degree. As an example, Isla de Lobos (not figured) between Tuxpan and Tampico receives 1848 mm., a much higher rainfall than would be expected for a coastal station at this latitude.

In addition to mean annual precipitation the diminished influence of tropical air circulation at higher latitudes is seen in lack of seasonal rainfall distribution. The Gulf Coast of central Veracruz and southern Tamaulipas has a rainfall regimen similar to the monsoon type of Asia, with pronounced alternation between winter dry (water deficit) and summer wet (water surplus) seasons. To the north in the drier Río Grande embayment the more equable distribution typical of most of continental North America is approached (Fig. 2). The nature of the tropical dry season in eastern Mexico is illustrated by the fact that Brownsville, with a mean annual precipitation of 950 mm. is actually slightly wetter six months of the year than Veracruz with more than twice this amount annually. In considering the entire coast from 18° to 28° N. lat., the monsoon effect is strongest in the vicinity of Tuxpan (Table II).

TABLE II

Percentage of Annual Rainfall During Wet Season on Gulf Coastal Area

Station	N. Latitude	Percentage of Rainfall, June-October
Corpus Christi	27° 41′	48
Brownsville	25° 54′	59
Victoria	23° 44′	62
Tampico	22° 13′	73
Tuxpan	20° 57′	90
Veracruz	19° 12′	86
Alvarado	18° 47′	81
Coatzacoalcos	18° 09′	66

Climate and vegetation. — Although the preceding account demonstrates no obvious single area in the Gulf lowlands that might comprise an absolute climatic limit to all tropical habitats, there are sections of the Gulf lowlands gradient in which critical climatic points are approached. The causal relationship between the northern limit of a plant formation and its environmental controls is generally unknown. Edaphic elements are often primary causes in controlling the local distribution of a formation approaching its climatic limits. Despite these limitations, it would still appear that Tropical Rainforest of the outer coastal plain finds its latitudinal limit between Coatzacoalcos and Alvaredo, in areas with less than 2400 mm. rainfall. Tropical Deciduous Forest, not found in Nuevo León by Muller, in the San Carlos area of Tamaulipas (Dice, 1937), or in the vicinity of Ciudad Victoria (see p. 31), terminates south of the Tropic of Cancer within the 1200 mm. isohyet. Here a complicated interdigitation with Thorn Forest and Thorn Scrub is partly under edaphic control and partly related to physiographic features and consequent orographic rainfall. The eastern foot and lower slopes of the Sierra San José de las Rusias, Sierra de Tamaulipas, and Sierra Madre Oriental are mantled with Tropical Deciduous Forest, whereas more arid Thorn Forest, Thorn Savannas, and Thorn Scrub occupy the lowlands between the Sierras.

Thus far I have considered only climatic change along an idealized coastal gradient, treating this as a model for similar latitudinal trends inland and in the Sierra Madre Oriental. Actually, conditions recorded at coastal stations are far from representative of interior localities at the same latitude and altitude. General trends in both vegetation and data from a few weather stations show that rainfall diminishes slightly from the coast toward the interior, rising again to a maximum in response to the orographic influence of the Sierra Madre. Shreve (1944) has diagramed the east-west gradient through northern Mexico; Goldman (1951: 258) commented on the increase in moisture through part of Tabasco from the vicinity of the Gulf to the foot of the Sierras of Chiapas. This effect results in the more humid lowland forests at any given latitude hugging the foot of the mountains whereas drier types are found toward the Gulf, a condition illustrated in a general way by Leopold's map (1950). In the Gómez Farías region Tropical Semi-Evergreen Forest is found at the very foot of the Sierra Madre and Thorn Forest and Savanna cover the middle of the coastal plain.

Climatic Gradients in the Gómez Farías Region

Only two weather stations in or immediately outside the Gómez Farías region are listed by Contreras (1942). One at Santa Elena, 161 m., south of Limón, is representative of the drier part of the lowlands; the other, Jaumave, 735 m., represents the dry interior valleys west of the Sierra Madre front. Weather conditions encountered for each in turn are as follows: mean annual temperature, 25.7° C, 21.2° C; mean of the coldest month, January, 19.9° C, 15.8° C; absolute minimum, -4° C, -4.5° C; precipitation 1080 mm., 568 mm.; percentage falling between June and October, 79, 76.

Four hygrothermograph stations were operated along the Sierra Madre front during part of the winter and spring of 1953. All were situated within 13 km. of each other. Although the records from these stations are too brief to characterize the climate of the region, they afford some data on altitudinal and habitat variation.

Four Bristol instruments, Model No. 4609TH, housed in louvered shelters with a double roof, were placed on the ground in the following sites:

1. Tropical Deciduous Forest (see p. 31), *ca.* 2 km. east-southeast of Pano Ayuctle at Rancho Cerro Alto, elev. 100 m. Although surrounded by trees and thorny shrubs the vegetation was in a dormant, leafless condition and the shelter received little shade until May.

2. Tropical Semi-Evergreen Forest (see p. 33), *ca.* 2 km. west-southwest of Pano Ayuctle at the north end of a spur of the Sierra Madre, elev. *ca.* 120 m. The shelter was located on a rocky slope partly shaded from the morning sun. The forest here was about 50 per cent deciduous during February-March and almost completely green by May.

3. Cloud Forest near Rancho del Cielo (see p. 34), elev. 1070 m., station operated continuously from February, 1953, to February, 1955, and continuing in operation. An adjacent clearing and recent lumbering partly opened the canopy, but the station was still heavily shaded.

4. Humid Pine-Oak Forest (see p. 36), *ca.* 7 km. northwest of Rancho del Cielo, elev. 1960 m. Trees here were evergreen but not spaced closely enough to provide a continuous canopy. The shelter was surrounded by a dense shrub growth.

Times of synchronous operation and means for this period are given in Table III. Calibration corrections were added for three instruments after field work ended.

TABLE III

Mean temperatures in degrees C recorded at four hygrothermograph stations in the Gómez Farías region, spring, 1951. The instrument at Station 3 was not calibrated.

Station	Meters Elev.	Instrument Calibration	Feb. 13-21	Mar. 9-25	May 5-13	May 25-June 6	Mean (47 days)
1	100	-1.3	22.5	29.1	30.9	32.7	29.1
2	120	+0.1	21.0	27.2	29.9	33.4	28.1
3	1080		15.0	20.6	21.4	23.5	(20.4)
4	1960	-1.5	10.4	15.0	19.0	17.8	15.6

The mean difference between Stations 1 and 2, 1^{0} C, reflects a difference in exposure. Midday temperatures in the more open, almost leafless, Tropical Deciduous Forest invariably exceeded those of the Semi-Evergreen Forest. At night, however, temperatures in both were virtually identical (Fig. 3). Both experienced nightly humidity rise, often to saturation; daytime drying out was much greater in the Deciduous Forest. The general daily cycle was quite similar at both stations.

Stations 2, 3, and 4 express altitudinal differences of *ca.* 1000 m. With regard to location these three were approximately equivalent, all benefiting from appreciable canopy insulation. By use of the data in Table III a lapse rate of about 0.8^{0} C per 100 m. between Stations 2 and 3 and 0.6^{0} C per 100 m. between Stations 3 and 4 was obtained. The time interval, 47 days, is too short to attempt an annual estimate, but it may characterize the dry season altitudinal gradient of the east-facing slopes in the Gómez Farías region.

Spring of 1953 produced a severe dry season, the summer rains not arriving until late June. During this period an unusually high frequency of hot, dry westerlies scorched the Chamal valley and other lowland areas. On five occasions hot, dry air, presumably of interior origin, overran the Sierra Madre and produced from one to six days of acute desiccation. The longest period was between February 28 and March 5. The effect was more pronounced in the mountains than on the coastal plain. Normally, Cloud Forest and Pine-Oak Forest humidities are high at night with a cyclic midday drop in clear weather, or continual high humidity (over 90 per cent saturated) in cloudy weather. During the period of the presumed westerlies no nocturnal rise in humidity took place, and the air remained less than 50 per cent saturated during the night as well as in the day. The two lowland stations were not affected in this manner. Although they registered slightly higher temperatures and lower midday humidity readings

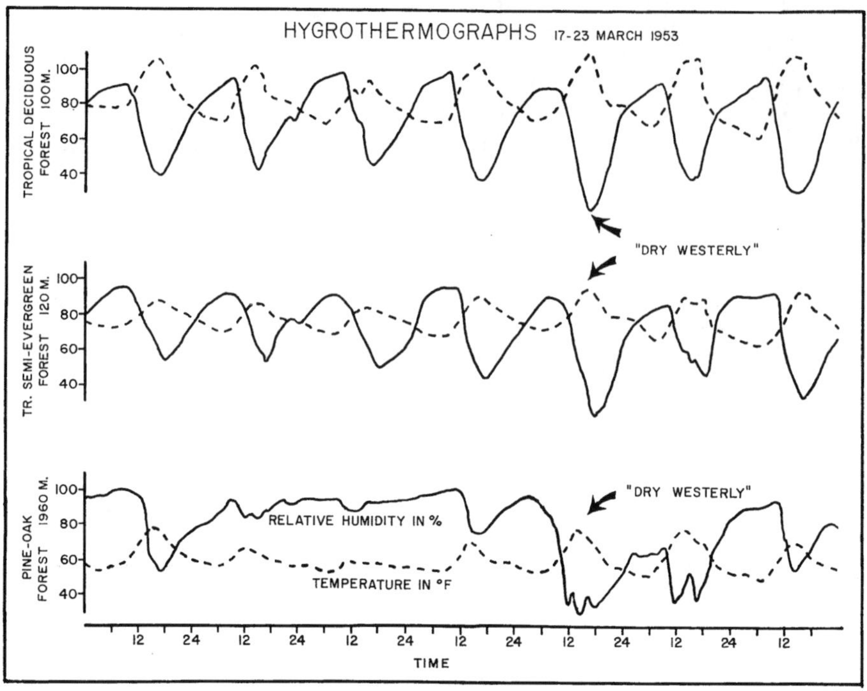

Fig. 3. Temperature and relative humidity curves from three stations in the Gómez Farías region, March 17-23, 1953. For simultaneous records at all four stations compare these with the middle chart of Figure 4. Arrows mark onset of a "dry westerly," which is felt more severely in the mountains.

than usual, the general effect was not pronounced (see Figs. 3 and 4). The dry westerlies of the mountains frequently arrived sometime after midnight bringing an abrupt drop in relative humidity and slight rise in temperature. Such an event in the nocturnal regimen is unknown under normal circumstances.

The chief difference between 1953 and 1954 with regard to the dry westerlies was their late occurrence in the dry year of 1953. In May of 1954 nearly constant temperatures and high humidities resembled those of June, July, or any other wet season month. In 1953 not only was the daily cycle quite pronounced but a period of very dry weather occurred (April 28 to May 2), a condition typical of a dry season month.

Cloud Forest Climate

Harrell (1951:29-36) treated general aspects of weather in the area, including hurricane frequencies and potential evaporation estimates. The following data serve to supplement his discussion. Two years of reasonably continuous hygrothermograph records and precipitation estimates represent the efforts of Frank Harrison. The thermal regimen in this

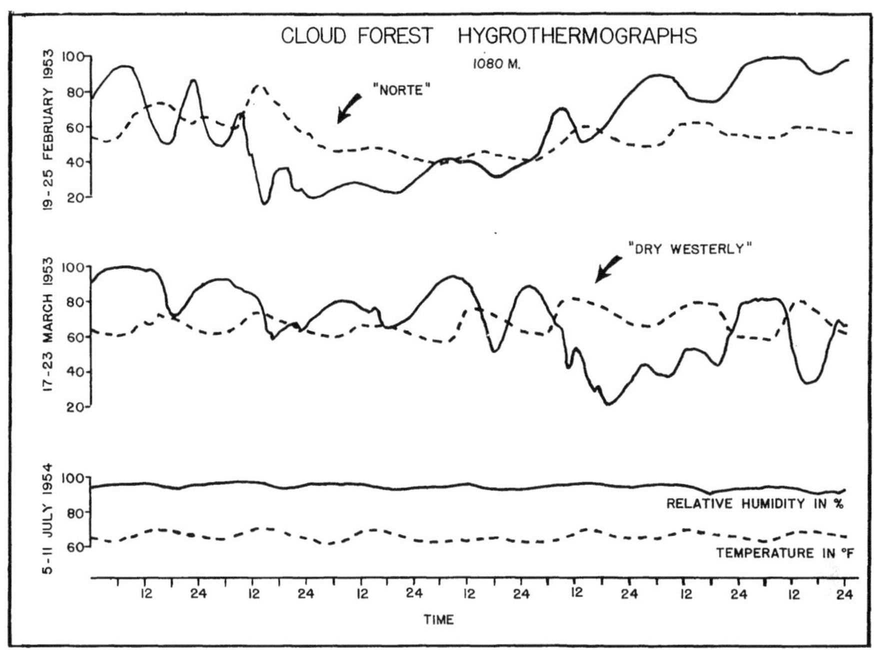

Fig. 4. Cloud Forest temperature and relative humidity curves during dry (February, March) and wet (July) seasons. Arrows mark onset of a "Norte" on February 20 and a "dry westerly" on March 21.

period, February 1953, to January, 1955, is illustrated in Figure 4. Mean annual temperature, averaged from four daily readings, 6:00, 12:00, 18:00, and 24:00 hours, was 19.4° C for 330 days in 1954. The ranges illustrated in Figure 5 represent weekly extremes averaged for the month. This procedure emphasizes the erratic changes typical of winter months in which a period of mild humid weather may be succeeded by hot, clear days, followed by arrival of cool north winds and low temperatures. Thus the annual regimen features a variable, heterogeneous season (dry) in which the annual thermal extremes occur, and there is a constant, homogeneous season (wet) which is warmer on the average but never attains the weekly maxima reached during the winter. Summer weather is controlled by the prevailing easterlies; winter, by weaker, drier, easterly trades, irregular polar outbreaks, and occasional dry westerlies.

Mr. Harrison also made rough measurements of rainfall by use of an open basin (Table IV). The seasonal difference between 1953 and 1954 is notable, especially the late arrival of summer rains in 1953. This had a disastrous effect on agriculture at La Joya de Salas. The villagers were unable to plant corn in time to mature a crop before the regular September frosts of the La Joya valley.

From the Rancho del Cielo records I estimate that the mean annual rainfall of the Cloud Forest lies between 2000 and 2500 mm., one of the

TABLE IV

Cloud Forest Rainfall in Inches. Figures from estimates of Frank Harrison.

Year	Jan.	Feb.	Mar.	Apr.	May	June	July	Aug.	Sept.	Oct.	Nov.	Dec.	Total	
													inches	mm.
1953	0	0	2	2.5	0.5	5.75	12.5	20.5	5.25	13.0	3.5	1.5	67	1700
1954	2.1	3.9	0	7.6	15.6	17.0	16.3	17.3	15.4	21.9	8.8	- -	125.9+	3200+

highest amounts received in northern Mexico, and certainly the heaviest fall at this altitude north of the Xilitla region of San Luis Potosí. It is roughly twice the amount received by the adjacent lowlands near Limón.

North of the Gómez Farías region the lowland rainfall diminishes from 1080 mm. near Limón to 900 mm. at Ciudad Victoria and 800 mm. at Linares in Nuevo León. Were it not for this decrease in the precipitation effectiveness of the summer trades, there is reason to believe that the mountains northwest of Victoria, which rise to over 2600 m., would

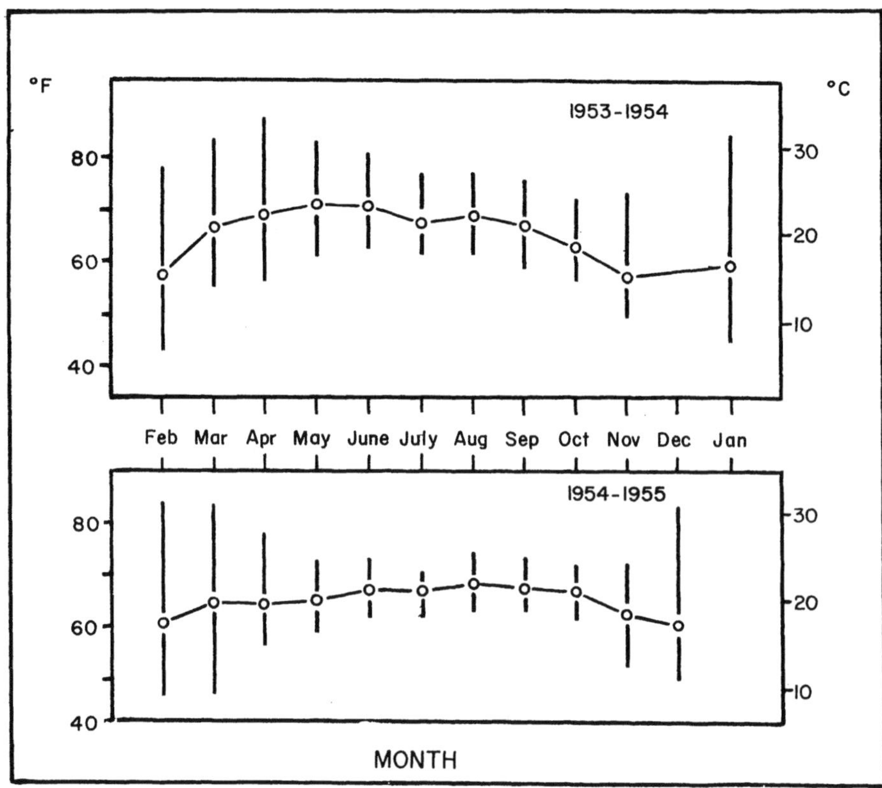

Fig. 5. Thermal regimen in Cloud Forest over a two-year period. Monthly means and ranges determined by method described in text. Note greater variation in dry season (Nov. to May) than in wet season (June to Oct.) ranges.

receive more rainfall than the Sierra de Guatemala. The more gentle rise of the escarpment near Victoria, offering less of an obstacle to the easterly trades, may also contribute to drier mountain climates in that region.

In the general problem of moisture availability the amount of cloud and fog insulating the mountain forests through the dry season is significant. Although no quantitative data were obtained, frequent observations show that the mountains above Gómez Farías are more often hung with clouds than those elsewhere in southern Tamaulipas. On numerous occasions the panorama of the Sierra de Guatemala as viewed from the sun-baked lowlands near Limón revealed dense clouds over the mountain forests (Pl. III, Fig. 2). On occasion, the entire front of the mountain from top to bottom would be cloud-bathed while the adjacent lowlands were clear and dry. Sutton and Pettingill (1942:4) commented: "...the Rancho [Rancho Rinconada on the Río Sabinas] will long be remembered for its wet, misty, cloud-hung weather." Their visit coincided with the height of the dry season when such days should have been at a minimum. In 1953 cloud periods were less frequent, but none the less evident.

Since no weather data are available on other climatic types of the Gómez Farías region, they can only be inferred from the nature of various vegetation types.

Summary

Although essentially tropical in climate, the Gómez Farías region is subject to severe continental outbreaks of polar air which may produce killing winter frosts. A belt of weakened but still effective trades brings to the lowlands an annual rainfall of between 1000 and 1400 mm., falling mainly in the months of May to October. The abrupt rise of the Sierra Madre Oriental from the coastal plain and its considerable frontal elevation in this area, 2400 m. at the highest point, produces a maximum of orographic rainfall. The montane forests receive more precipitation than any other area in eastern Mexico below 1600 m. elevation and north of latitude $21^{\circ} 30'$. Estimates from the Cloud Forest (1070 m.) place this at over 2000 mm. annually. Equally important, dry season insulation through clouds and fogs also reduces evaporation along the mountain front.

The mean annual temperature in the lowlands is about 25° C, and a dry season lapse rate of 0.6° to 0.8° C per 100 m. is estimated on the eastern side of the Sierra Madre. The effect of this gradient and the varying, but generally heavy, precipitation on easterly slopes in contrast with reduced precipitation and greater evaporation on westerly slopes produces a variety of climatic types in the mountains. These areas lack weather stations, are usually not mapped in climatic atlases, and can be recognized best in terms of natural vegetation.

Vegetation

The preceding geological and climatic features interact with certain biotic factors to determine the vegetation of the Gómez Farías region. In the following description my purpose is threefold: (1) to outline in terms of structure, function, and dominant flora the natural, ostensibly climax, vegetation types of the Gómez Farías region; (2) to discuss zonal and ecological behavior of a few conspicuous species that may serve as plant indicators; and (3) to attempt a correlation of vegetation types in the Gómez Farías region with others described elsewhere in Mexico. The observed relationship of fauna to vegetation and a presentation of certain historical problems will be treated subsequently.

Vegetation Types in the Gómez Farías Region

The level of ecological abstraction with which I am chiefly concerned is the Plant Formation (Schimper, 1903) or Vegetation Zone (Leopold, 1950). These terms will be used interchangeably in the following discussion. At lower levels of integration the associations, synusia, and biotypes recognizable within each formation are given no more than cursory treatment. The sequence of forest types along the coastal plain and into the Sierra, as illustrated in Figure 6, is greatly simplified. Many factors — topographic, climatic, edaphic, biotic, and cultural — produce an exceedingly complex vegetational pattern. In dividing the area into eight formations I do not deny the existence of broad transitional areas. The entire sequence across the coastal plain from arid Thorn Scrub of the middle Coastal Plain to Tropical Semi-Evergreen Forest at the foot of the Sierra Madre might be considered a transitional belt or continuum by some ecologists. In the Sierra Madre formations are usually more obvious and their boundaries sharper, but even here absolute distinctions cannot be made. In the absence of a detailed study of vegetation it is convenient to establish arbitrary divisions, while acknowledging, if not describing, the ecotones (transitions).

Figure 6 and Map 2 emphasize three important points: (1) the isolated position of humid montane forests, Humid Pine-Oak Forest, Cloud Forest, and Tropical Semi-Evergreen Forest, in the Gómez Farías region; (2) the absence of Tropical Deciduous Forest; Tropical Semi-Evergreen Forest, and Cloud Forest north of the Gómez Farías region; and (3) the wealth of vegetation types in this small area, which includes eight of the twelve zones recognized in Mexico by Leopold (1950:508).

Nomenclature in the following list essentially follows that of Standley, *Trees and Shrubs of Mexico*, 1920-1926. I sought to collect only dominant plants in each habitat, especially trees and shrubs; many of these, however, Standley has not recorded from Tamaulipas, a fact which illustrates how poorly the flora is known. Many other additions to the Tamaulipan flora have been reported by Harrell (1951), Sharp *et al.* (1950), Sharp (1954), and others collecting in the Cloud Forest. In addition to these the following list includes at least 33 new state records. For most of the identifications I am indebted to Rogers McVaugh. Oaks were identified by C. H. Muller. Specimens representing approximately 210 numbers were deposited in the University of Michigan Herbarium. Field numbers of these are enclosed in parentheses. A few other names are based on literature records or field observations.

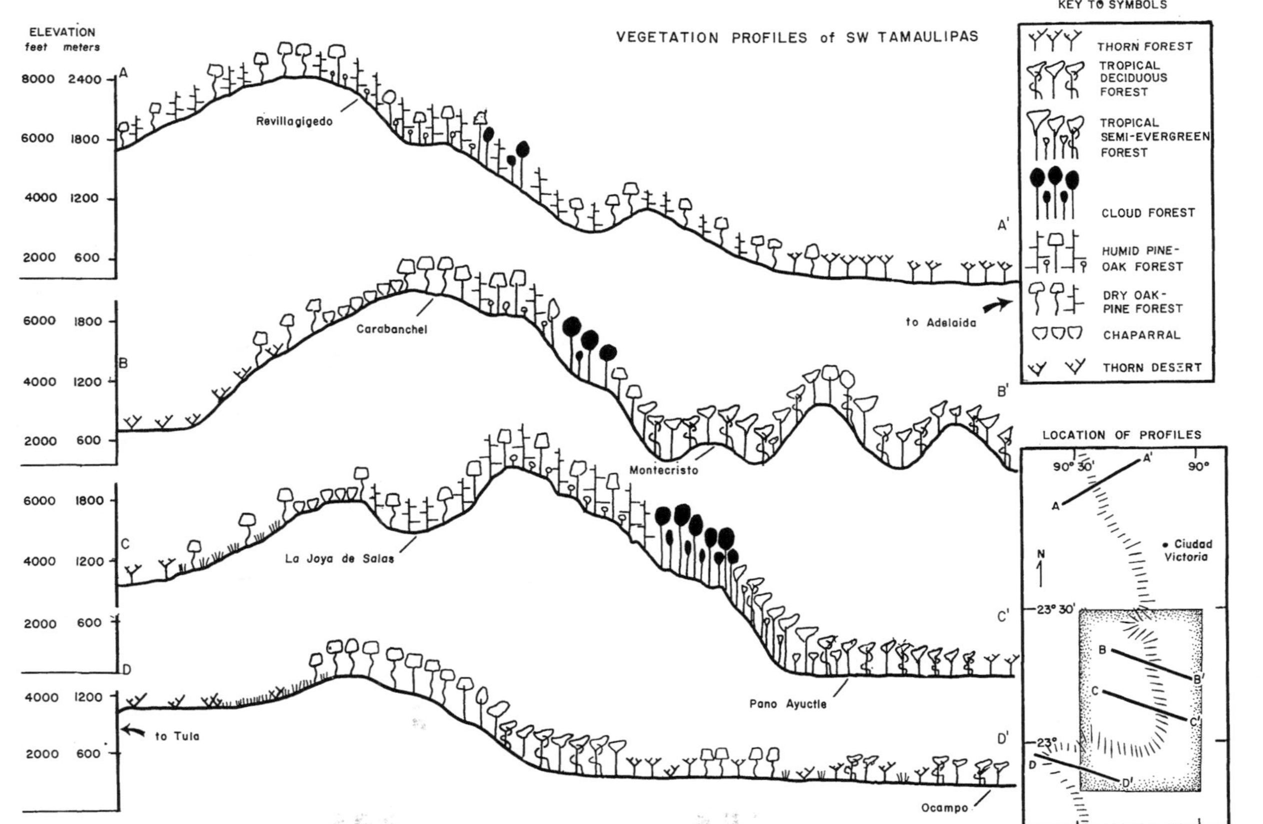

Fig. 6. Vegetation profiles across the Sierra Madre Oriental of southwestern Tamaulipas. The shaded part of the lower inset marks the Gómez Farías region as shown in Maps 1 and 2. **Plates I to VII** illustrate each of the formations found across the Sierra Madre along profile C - C′.

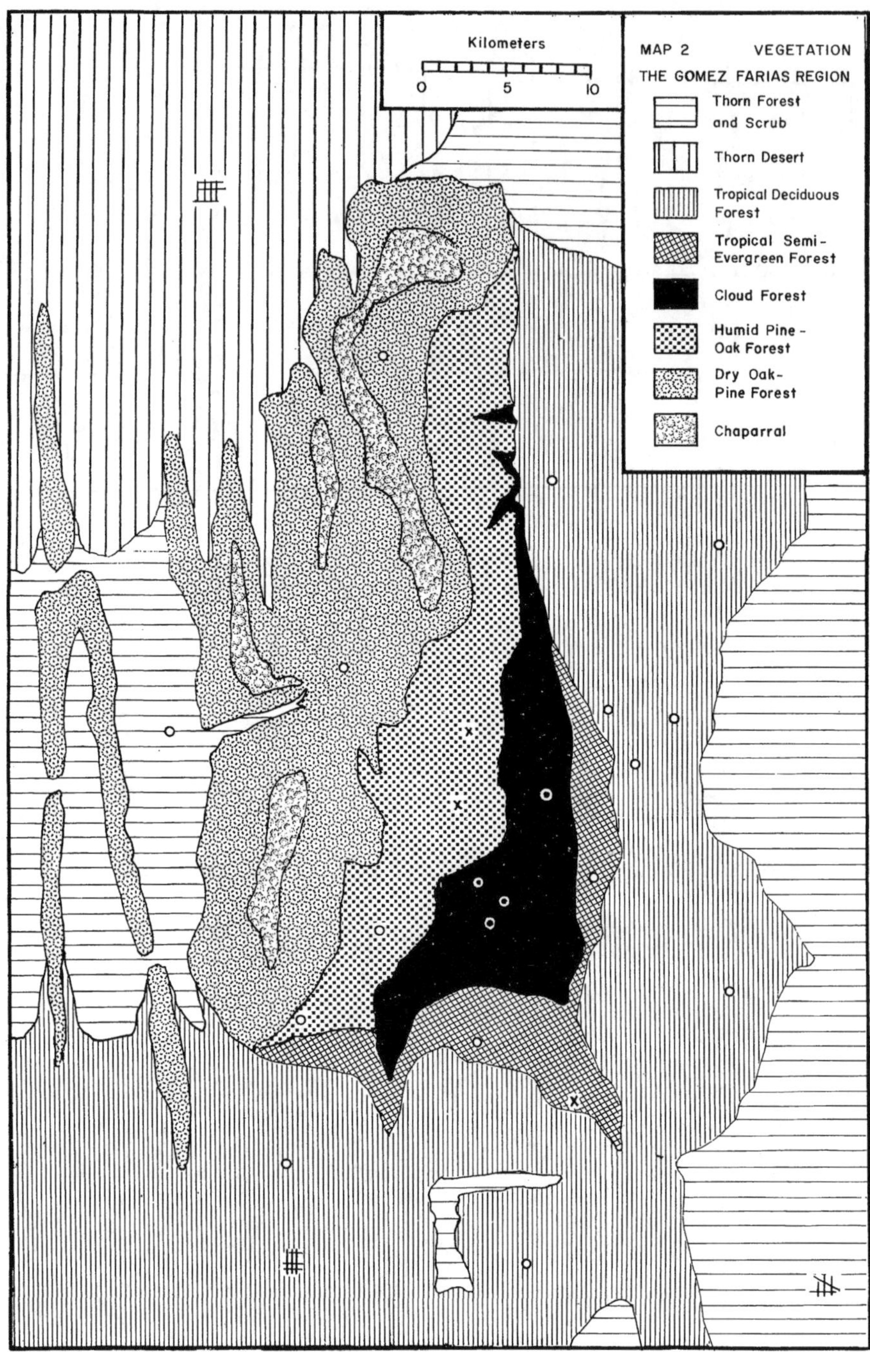

Map. 2. Natural vegetation of the Gómez Farías region, 22° 48′ to 23° 30′ N. lat. and 99° to 99° 30′ W. long. All the localities figured on Map 1 are indicated. For specific locality names see Map 1.

The vegetation types recognized are: (1) Thorn Forest and Thorn Scrub, (2) Thorn Desert, (3) Tropical Deciduous Forest, (4) Tropical Evergreen and Semi-Evergreen Forest, (5) Cloud Forest, (6) Humid Pine-Oak Forest, (7) Dry Oak-Pine Woodland, and (8) Montane Chaparral.

A description of some typical examples and certain modifications of these types follows.

Thorn Forest and Scrub (Pl. I, Fig. 1). — Under this heading are included many dry lowland and interior plant associations, all characterized by low trees and shrubs, usually thorny and deciduous, and either microphyllous or compound-leaved. A variety of factors — climatic, edaphic, and cultural — may be responsible for the development of either dense Thorn Forest, lower, more open Thorn Scrub, or Thorn Savanna. I have not attempted to untangle these. The biological and climatic changes encountered in this formation between southern Texas and southern Tamaulipas deserve special study. Originally this section of the Gulf Coastal Plain may have been covered with extensive grassland. A free translation of the account by Santa María (p. 369) suggests invasion of Thorn Scrub after Spanish conquest: "...the arable land had no useless thorny shrubs to spoil its natural abundance. Since the arrival of the white man there has been a plague that has injured and converted into horrible form that which was previously beautiful. Already there are innumerable, spiny, most pernicious shrubs."

Within the Gómez Farías region along the Mexico-Laredo highway immediately south of Llera is a dense Thorn Forest. Various trees, predominately deciduous and occasionally reaching a height of 6 m., included the following: *Acacia coulteri* (PSM 096), *Lantana involucrata* (PSM 099), *Caesalpinia mexicana*, *Cordia boissieri* (PSM 098), *Neopringlea integrifolia* (PSM 097), *Pithecolobium* sp., and *Yucca* sp.

About 30 km. east of Llera, outside the Gómez Farías region, an uninhabited rolling plain at the foot of the Sierra de Tamaulipas is covered with a dense grass sward. Scattered throughout are yuccas and a tree, *Piscidia communis* (PSM 101), also common in Tropical Deciduous Forest. Possibly this savanna represents a remnant of the original pre-Columbian lowland vegetation.

The natural vegetation about Limón may originally have included extensive grassland. Before irrigation and sugar-cane cultivation much of this district was what local residents describe as "brush country." I interpret this as mainly Thorn Forest; however, Everts Storms, a resident in the region since 1910, related that older inhabitants had informed him that the area about Mante and Limón once was grass-covered. Mesa tops east of Pano Ayuctle were, and in some places still are, mainly grass and low tree savanna. According to Seymour Taylor, a resident in the Chamal valley since before the Mexican revolution, his ranch northeast of Chamal was formerly "tall prairie." Conceivably, grazing and cultivation promoted the spread of palms and thorny trees into the area.

Additional clues to the past vegetation of the Gómez Farías region appear in the diaries of early travelers. Proceeding from Tula toward Tampico in December of 1822, Poinsett (1825:262) described Santa Bárbara (= Ocampo) as "...surrounded by a variety of beautiful evergreens, oranges, bananas, and mimosas of great height, and some more than fifteen feet in circumference." The Chamal valley was covered with palm trees, then as now. Poinsett found the eastern foot of the Sierra Cucharras heavily wooded (probably Tropical Deciduous Forest). Near Limón he "passed through a more open country, interspersed with cultivated fields and trees of mimosa, yucca, and palms." At the Río Limón (= Río Guayalejo ?) he noted lofty mimosas five to six feet in diameter with bamboo ("canes") thirty to forty feet high. Continuing eastward, "On leaving the margin of the river we left all appearance of rich and luxuriant vegetation and for six hours passed over a plain, arid, parched, and thinly wooded with mimosas and small shrubs."

Another visitor in the region traveling westward emphasized a difference between the Ocampo (Santa Bárbara) Valley and the country to the east (Lyon, 1828:130).

May 21, 1826 - "In this vale [Ocampo] I saw for the first time, in Mexico, bright green fresh-looking herbage, as verdant as that of our English fields. Nothing could be more striking than the change perceptible in one morning's ride over the mountains, —on the other side of which, the whole way to the sea coast, the grasses were the color of blighted corn."

Today no such striking difference distinguishes the Ocampo area which appears dominated by Thorn Forest similar to that of the coastal plain. Also of significance in Lyon's account is the implication that grass was a dominant feature of the landscape east of the Ocampo Valley.

From the accounts of Poinsett and Lyon I conclude that despite its present overgrazed state and resemblance to Thorn Forest the Ocampo Valley was once largely Tropical Deciduous Forest. Evidently palm bottom is a long-enduring forest type in the Chamal valley. The natural landscape east of the Sierra de Cucharras is more difficult to interpret; however, both Poinsett and Lyon confirm the view that Thorn Forest and savanna were present, with heavy gallery forest along the rivers.

On Méndez shale near the source of the Río Frío in an area surrounded by Tropical Deciduous Forest grow a variety of xeric thorny species including organ pipe cactus, *Opuntia*, *Yucca*, *Acacia amentacea* (PSM 005), huisache, and other species that suggest Thorn Forest. This is one of several such areas; others occur in the hills around Chamal. All are probably under edaphic control.

West of the Sierra de Guatemala runs a series of at least five narrow ridges separated by eroded, rather narrow valleys. Two of these valleys lie in the Gómez Farías region and the one I visited near San Antonio supported Thorn Forest and Thorn Savanna. Trees in this area were not densely spaced, generally not so thorny, and somewhat taller (one *Bumelia* along a dry arroyo was 11 m.) than those of the coastal plain. Yuccas, *Opuntia stenopetala*, small agaves, and *Cordia* were seen here, and the following were collected: *Acacia amentacea* (PSM H51), *Bumelia laetevirens* (PSM H19), *Morkillia mexicana* (PSM H20), and *Rhus terebinthifolia* (PSM H21). According to present inhabitants the area supported a larger settlement in the past when corn and other crops were cultivated. At present, grazing is the chief agricultural activity, and the two families living at San Antonio raise little besides bees, nopal cacti, and cattle. The ranch of San José about 1 km. south of San Antonio is abandoned.

To the north of San Antonio the valley rises slightly, probably not exceeding 1100 m., and the Thorn Forest merges with a drier vegetation type, Thorn Desert. The valley thus comprises a dry corridor completely isolating various humid and subhumid montane forests in the Sierra de Guatemala (see Map 2).

Thorn Desert (Pl. VII, Fig. 2). — Lying in the rain shadow, behind the Sierra Madre front, the valley of Jaumave, at 730 m., is the driest part of the Gómez Farías region (560 mm. annual rainfall). The unusual climatic combination of a very dry valley, drier than most of the coastal plain, and low elevation, warmer than most of the Central Plateau, produces a distinctive vegetation. Although unusually dry for eastern Mexico, the Jaumave Valley is much moister than desert areas on the Pacific slope. Many species grow here that are not found elsewhere in the Gómez Farías region. The valley is well known for its variety of Cactaceae. A partial inventory of the flora by Bravo (1952) included the shrubs *Jatropha spathulata, Prosopsis chilensis, Leucophyllum texanum, Coldenia canescens, Cercidium floridanum, Koeberlinia spinosa, Agave funkiana, A. lecheguilla,* and *Yucca treculeana.* Of the cacti, *Opuntia leptocaulis, O. pumila, O. kleiniae, O. imbricata,* and *O. stenopetala* are among the 17 species reported. The valley combines an interesting mixture of Central Plateau desert plants (*Opuntia imbricata, O. stenopetala*) with lowland Thorn Scrub species (*Leucophyllum, Cercidium,* etc.). As an animal environment it is unusual in eastern Mexico, combining a sparse rainfall with a high mean annual temperature.

Tropical Deciduous Forest (Pl. I, Fig. 2; Pl. II, Figs. 1 and 2). — In the lowlands near the foot of the Sierra Madre and on the lower limestone ridges and slopes, this formation is widespread. It includes most of the areas devoted to lowland tropical agriculture. North of latitude 23° 30′ increasing aridity and possibly also increasing frost frequency limit Tropical Deciduous Forest in its northward extent. In ravines west of Ciudad Victoria some of the floristic elements of Tropical Deciduous Forest are present near water courses, but the Bombacaceae, Burseraceae, *Guazuma ulmifolia,* and *Beaucarnea,* typical of this formation farther south, are conspicuously absent. On slopes the vegetation is exclusively Thorn Forest, gradually transitional to oak-pine forest above 500 m. Muller (1937, 1939) and White (1942) did not find Tropical Deciduous Forest in Nuevo León.

Typically, the Tropical Deciduous Forest is formed of trees of medium height (12-15 m.) which are rather widely spaced and rise out of a dense, almost impenetrable, understory of lower trees, about 5 m. in height. The forest is leafless in winter; in 1953, new leaves appeared in late March and April, at least two months ahead of heavy rains in late June. The months of April and May of that year were uniformly hot and dry. In this period severe wilting was evident. Lianas and small tillandsias are common; epiphytic orchids are present. A characteristic shrub throughout lowland forests is the formidably spined *Bromelia pinguin,* which may also grow in areas of Thorn Forest. Extreme density at and near ground level is characteristic of Tropical Deciduous Forest. The habitat is dominated by phanerophytes. Most of the trees have medium-sized leaves with many compound-leaved species present. An arborescent *Opuntia* and the lianoid cactus *Acanthocereus pentagonus* are characteristic. At hygrothermograph station 1 (*ca.* 2 km. east southeast of Pano Ayuctle) the following were collected growing on hard-packed black earth: *Croton cortesianus* (PSM 079), *Lasiacis divaricata* (PSM 077), ?*Marsdenia coulteri* (PSM 095), ?*Schoepfia* sp. (PSM 076). The genera *Ficus, Acanthocereus, Enterolobium, Bromelia,* and *Cassia* also occur here.

Tropical Deciduous Forest on coarse limestone ridges such as the Sierra Cucharas, Sierra de Chamal, and foothills of the Sierra Madre includes species seldom found elsewhere. In these areas the bulbous trunk of *Beaucarnea inermis* lends a distinctive, bizarre aspect to the landscape. Other species collected on the ridges near Chamal and Gómez Farías include: *Acacia coulteri* (PSM 071c), *Bactris* sp. (PSM 015), *Begonia heracleifolia* (herb) (PSM 012), *Bombax ellipticum* (PSM 016), *Cassia emarginata* (PSM 071b), *Croton niveus* (PSM 072), *Kalanchöe pinnatum* (introduced, PSM 030), *Petrea arborea* (PSM 031), *Piscidia communis* (PSM 111), and *Pisonia aculeata* (PSM 032).

A common tree (Burseraceae) with a smooth trunk and orange-colored bark, locally called "chaca," was not identified as to species. A frequent species along roadsides, and in some undisturbed conditions north of Chamal, is *Guazuma ulmifolia* (PSM 119), locally called "aquiche."

Gallery forest, composed of tall trees, largely evergreen including cypress (*Taxodium mucronatum*), *Inga spuria* (PSM 081), *Ficus segoviae, Platanus* sp., *Salix humboldtiana, Guadua* sp. (Darnell, 1953), grows along rivers. The gallery forest forms an evergreen ribbon running through both Tropical Deciduous Forest and Thorn Forest. In headwater areas, where surrounded by Tropical Semi-Evergreen Forest, the gallery forest loses much of its structural distinctiveness.

Another evergreen forest found in areas of Tropical Deciduous Forest is palm bottom, a dense consociation of *Sabal* that may exceed 25 m. in height. Many of the extensive palm bottoms of the Chamal and Sabinas valleys have been cleared for cultivation; others are burned seasonally. Palm bottom is found mainly in low-lying areas over loose black alluvium near rivers, at greater distance from the water than gallery forest. Isolated palm trees may appear on foothills and slopes above the bottoms.

A third variant of the Tropical Deciduous Forest, possibly under edaphic control, is a type of oak woodland. Scattered live oaks (*Quercus oleoides*) are found near Encino at 100 m. In the lowlands west and north of Ocampo at 400 m. there are groves of this species (PSM 115, 047) with an understory of *Acacia pennatula* (PSM 116) and various grasses. The oaks reach 13-15 m. in height, are evergreen, and may support a variety of small epiphytes including tillandsias, orchids, and *Rhipsalis* sp. Another species of oak which, unlike *Q. oleoides*, is not confined to low elevations is *Q. polymorpha* (PSM 118). North of El Tigre, elevation 500 m., in a region surrounded by cultivated fields and Tropical Deciduous Forest, a parklike copse of this oak grows to a height of 13 m. in a stand of tall grass, *Arundinella deppeana* (PSM 120).

Where Tropical Evergreen Forest does not intervene, Tropical

Deciduous Forest reaches an elevation of 800 to 900 m. before being replaced by dry oak woods. The trail between Ocampo and Tula traverses such an area. Ordinarily in the Gómez Farías region Tropical Deciduous Forest does not exceed 600 m.; above this point it is replaced by the following type.

Tropical Semi-Evergreen and Evergreen Forest (Pl. III, Fig. 1). — From the coastal plain along an idealized transect to the Sierra Madre, the following five changes in vegetation are outstanding: trees increase in average height from 3 to 25 m.; the percentage of evergreen species increases from less than 10 to over 50; a preponderance of microphylls (small-leaved species) is replaced by mesophylls with a few megaphylls; the percentage of thorny species decreases from more than 70 to less than five; and finally the number of lianas and large epiphytes increases greatly. Whether this change represents a continuum, or a series of discrete, absolute changes in which areas of homogeneity exceed those of transition, I am uncertain.

At the humid extreme of this gradient, controlled by orographic rainfall, is found a forest dominated by tall evergreen or semi-evergreen trees reaching a height of 25 m. Such forest is best developed north of Chamal at Aserradero del Paraíso. This is the only locality in the Gómez Farías region where intensive lumbering of tropical woods is feasible. Cedro (*Cedrela* sp.), *Enterolobium* sp., *Quercus germana*, and "Palo Santo" (probably *Dendropanax* sp.) are some of the species exploited. Forest in this area is tall and open enough so that one may walk through it without difficulty, unlike much of the Tropical Deciduous Forest. Elsewhere in the Gómez Farías region the Tropical Semi-Evergreen Forest may be quite low in height, as north of El Tigre and on exposed slopes along the Sierra Madre Front. Tall forest is found only in ravines or on level ground where at least a shallow soil can accumulate. The areas of deeper soil, such as those about Pano Ayuctle, have been largely cleared and are cultivated in sugar cane. Between Gómez Farías and Montecristo excellent sugar cane is cultivated, without benefit of irrigation, on soils which probably once supported heavy forest.

Species collected or observed in Tropical Evergreen and Semi-Evergreen forest include a variety of trees: *Abutilon* sp. (PSM 113), *Achatocarpus mexicanus* (PSM H11), *Brosimum alicastrum* (PSM 088), *Celtis monoica* (PSM 087), *Dendropanax arboreus* (PSM 112), *Enterolobium* sp., *Ficus* sp. (strangling fig), *Gymnanthes actinostemoides* (PSM 065), *Iresine tomentella* (PSM 123), *Quercus germana* (PSM 124), ?*Spondias* ("jobo"), *Tabernaemontana citrifolia* (PSM 066), *Ungnadia speciosa* (PSM 064), *Viburnum* sp. (PSM 067).

The following common herbs, shrubs, and lianas were collected in this formation: *Acalypha schlechtendaliana* (PSM 082), *Bauhinia mexicana* (PSM 085), *Campelia zanonia* (PSM 022), *Epidendrum cochleatum* (PSM 083), *Heliconia* sp. (PSM 017), *Hybanthus mexicanus* (PSM 086), *Randia laetevirens* (PSM 084), *Rhipsalis cassutha* (PSM 021), *Setaria poiretiana* (PSM 020), *Solanum lanceifolium* (PSM 063), *Zamia* sp.

I am unaware of any formations similar to this one north of the Gómez Farías region. Muller (1937, 1939), White (1942), and others found no equivalent in Nuevo León. Unlike Tropical Deciduous Forest, Tropical Semi-Evergreen Forest is not continuously distributed south of the Gómez Farías region, and the formation in this area may represent a relic isolated at its northern limit. The nearest forest to the south that may be equivalent is that west of the Río Naranjo near Naranjos in San Luis Potosí, *ca.* 40 km. south southwest of the Gómez Farías region.

Low evergreen thickets of *Zamia, Beloperone,* and *Acalypha schiedeana* (PSM 114) with scattered trees including *Quercus germana* and *Dendropanax* are found north of El Tigre at 800 to 1000 m. Their classification is difficult. In growth habit the low, dense nature of this area and the scarcity of herbaceous bromeliads other than Spanish moss suggested the structure of Tropical Deciduous Forest. The trees and shrubs, however, were in full leaf when visited in mid-April of 1953, and Tropical Deciduous Forest indicators, such as *Guazuma* and "chaca" (Burseraceae), do not reach this elevation. I suspect that fire may have modified much of the area. This, in addition to the fact that it lies close to the arid rain shadow valleys west of the Sierra Madre front, may account for its poor development as Evergreen Forest.

Cloud Forest (Pls. IV and V). — Harrell (1951) provides a detailed description of Tamaulipan Cloud Forest. In terms of height, layers, and spacing this is the "optimal" vegetation type of the Gómez Farías region. Although Tropical Semi-Evergreen Forest and Humid Pine-Oak Forest may approach the Cloud Forest in their development they appear to lack the total density (biomass) of this formation. Supplementary descriptions, mainly of the flora, can be found in the following: Sharp *et al.* (1950), Hernández *et al.* (1951), Sharp (1954), and Martin (1955b).

Briefly the formation can be characterized as a tall (20 to 30 m., occasional trees to 40 m.), dense, semi-evergreen forest. The following are deciduous in midwinter: *Acer, Liquidambar, Cercis,* and *Carya.* New leaves are fully formed between mid-March and the first of April before the dry season ends. The forest is richly supplied with epiphytes, both large tank bromeliads and smaller ferns, orchids, mosses, and hepatics. A thick, spongy mat of mosses covers many of the large limestone rocks. Herbs are scarce under the tight canopy, which is seldom broken except near rock ridges and around karst sinks and lapies. Forest of this type may be encountered between 900 and 1700 m.; however, typical areas are confined to the interval between 1000 and 1500 m.

Among the dominant species of trees are *Quercus sartorii* (PSM 057), *Q. germana, Liquidambar styraciflua, Prunus serotina* ssp. *serotina* (PSM 024), *Podocarpus reichei* (PSM H35), *Magnolia schiedeana, Clethra macrocarpa,* and *Meliosma alba.* Some of the flora is shared with adjacent vegetation zones; *Chamaedorea, Zamia,* and *Quercus germana* occur also in Tropical Evergreen Forest, whereas *Carya, Magnolia, Ternstroemia,* and *Carpinus* are found in Humid Pine-Oak Forest. Many species, however, are confined to the Cloud Forest as defined in a structural sense. *Meliosma alba, Liquidambar, Podocarpus, Acer skutchii* (PSM H34), and *Quercus sartorii* are among the most reliable Cloud Forest indicators.

Within the formation several associations or segregates can be recognized. At lower elevations around 1050 m., *Quercus sartorii, Q. germana,*

Clethra macrophylla, and *Liquidambar styraciflua* are among the dominants. At higher elevation almost pure stands of beech (*Fagus mexicana,* PSM H27) with a characteristic undergrowth of *Illicium* occur locally near Rancho Viejo. Along the trail to Lagua Zarca at 1450 m., fir (*Abies*) and *Taxus* are mixed with *Liquidambar, Magnolia,* and various oaks. In exposed areas wherever pinnacles or karst lapies occur, the forest is replaced with a low dense tangle of shrubs, low trees, and a rather soft-leaved agave.

In 1953 Harrell and I sought to determine the northern limit of Cloud Forest. An area described by David J. Rogers (personal correspondence to A. J. Sharp, 1950) along the road between Adelaida and Dulces Nombres, northwest of Ciudad Victoria, included several species such as *Liquidambar* and *Magnolia,* which suggested Cloud Forest conditions. On a visit in February, 1953, we found these trees and *Meliosma alba* growing on a north-facing slope at about 1500 m.; however, they did not form a forest of the Rancho del Cielo type. The sweet gum trees were of low or medium height, and many other species which might be expected, such as *Podocarpus, Acer, Quercus sartorii,* and *Fagus,* were not found. The area may be considered Humid Pine-Oak Forest with a slight admixture of Cloud Forest species. Muller (1939) and White (1942) did not list *Liquidambar* or any of the other typical Cloud Forest species in Nuevo León. I am indebted to Ing. Marcelino Castañeda y Nuñez de Caceras, who knows Tamaulipan vegetation intimately, for the information that *Liquidambar* ("alamillo") is not a large tree when found in the mountains near Ciudad Victoria. He reported no tall dense forest with *Liquidambar* in Tamaulipas north of the Gómez Farías region.

Tracing the Cloud Forest north from Rancho del Cielo, Harrell and I discovered a remnant in two hanging valleys west of Montecristo. Here between 1150 and 1400 m. we encountered *Quercus sartorii, Meliosma alba, Liquidambar, Podocarpus, Quercus germana, Turpinia occidentalis* (PSM 135), and *Arisaema.* The forest was tall but less dense in its lower layers than the Rancho del Cielo type. Local residents of Montecristo reported alamillo in another valley, "Cañon del Diablo," slightly to the north. This, presumably, is the northern limit of *Liquidambar* in the region. We entered this valley from above, but were prevented from reaching the elevation where Cloud Forest might be expected by a sheer cliff at about 1700 m. The canyon above this point held fir (*Abies*), *Carpinus,* and the bottom was covered with nettles (*Laportea canadensis,* PSM H32).

Between Llera and Ciudad Victoria the Sierra Madre front is largely uninvestigated biologically; however, the mountain front here is not so high, seldom cloud-hung, and at a distance appears much drier than the section west of Gómez Farías. Present knowledge would indicate that Cloud Forest as a structural entity is not found north of the Gómez Farías region.

With its large number of tall, slender, often buttressed trees, its high percentage of deciduous or semi-evergreen individuals in the upper layers, and its strong floristic affinity with Eastern Deciduous Forest, Tamaulipan Cloud Forest is distinctly different from typical Montane Rainforest in the sense of Richards (1952) or Beard (1955). Historical connection with Eastern Deciduous Forest possibly accounts for some of the structural peculiarities, including a deciduous tendency not evident in typical Montane Rain Forest. Nevertheless, the unusual height and density may reflect more nearly optimal climatic conditions for tree growth in this locality than in Montane Rainforest elsewhere.

Between 1400 and 1700 m. Cloud Forest reaches its upper altitudinal limit. The boundary between Cloud Forest and the next adjacent type, Pine-Oak Forest, is less often characterized by a gradual transition than it is by interdigitation or by isolated pockets of one type within the other. The interdigitation is under physiographic control. The mountainside on which Cloud Forest and Pine-Oak Forest occur is highly porous with many karst features. In the bottom of Lagua Zarca and other large dolines which trap soil, a form of upper Cloud Forest exists in which tall *Liquidambar, Magnolia, Abies, Taxus, Podocarpus,* and many shrubs grow surrounded on all sides by open Pine-Oak Forest. Steep slopes as low as 1200 m. may be forested with pine and oak, whereas level benches and flats, in addition to the doline bottoms, are covered by Cloud Forest.

Humid Pine-Oak Forest (Pl. VI). — The distinction between the preceeding and the present type is structural as well as floristic. In the Humid Pine-Oak Forest sclerophylls and needle-leaved trees dominate in place of the mesophylls of the Cloud Forest. In spacing and layering there is a distinction. Cloud Forest as here defined is quite dense with a closed canopy 20 to 30 m. in height and a wealth of understory trees and tall shrubs. This density is not equaled in the Pine-Oak Forest, which is more open, even on level sites. In only one regard, density of low shrubs and herbs, does Humid Pine-Oak Forest exceed Cloud Forest. Evergreen species are more frequent in the canopy of the former. Other structural features of Pine-Oak Forest include: (1) tall, straight canopy trees, chiefly pine, with *Abies* and *Cupressus* appearing in moister places. The canopy trees average 25 to 30 m. in height with occasional individuals exceeding 35 m. (2) a characteristic lower level of sclerophyll oaks, *Arbutus,* and occasional *Magnolia* and *Carya* average 15 to 20 m. with a great variety of low and medium evergreen shrubs beneath. (3) many herbaceous monocotyledons predominate at ground level, but true grasses (Gramineae) are scarce. (4) epiphytes, notably mosses and hepatics, may be less evident than in Cloud Forest; lianas are virtually nonexistent.

The flora is rich and diverse, and the following list is far from complete.

Along the trail from Lagua Zarca to La Joya de Salas near hygrothermograph station 4 (1960 m. elevation) *Pinus montezumae* and *P. patula* are dominants along with *Quercus affinis* (PSM H7), *Q. diversifolia* (PSM H6), *Q. rugosa* (PSM 071a), *Arbutus ?xalapensis* (PSM H10), and *Cornus disciflora* (PSM 060).

An unidentified but distinctive shrub or low tree (PSM H37) having the shape of a *Yucca* but with drooping rather than rigid, erect leaves, is common and characteristic. Other shrubs are: *Viburnum elatum* (PSM H26), *Litsea muelleri* (PSM 045), and *Ternstroemia sylvatica* (PSM 054). The common herbs include *Manfreda guttata* (PSM 132), *Orthrosanthus chimboracensis* (PSM 137), *Spiranthes* sp? (PSM 061), and *Verbena elegans* (PSM 055).

Elsewhere in more humid ravines and level basins such as the Agua Linda uvala are found fir (*Abies*), yew (*Taxus*), *Cupressus benthami* (PSM 044), *Magnolia*, and *Carpinus*. *Tilia* (PSM H39) and *Carya* are uncommon at lower elevations near the lower limit of Pine-Oak Forest. In a few flat open places above 1700 m., tiny wet meadows support *Myrica mexicana* (PSM 073), *Castilleja* sp. (PSM H31), *Oxalis* sp. (PSM H41), *Ranunculus peruvianus* (PSM H42), *Zephyranthes* sp. (PSM H45), and many low sedges. Among the shrubs collected in the area mainly above 1700 m. are *Berberis lanceolata* (PSM H22), *Cornus excelsa* (PSM H14), *Ceanothus coeruleus* (PSM H17), *Garrya ovata* (PSM H12), *G. glaberrima* (PSM 125), *Phoebe effusa* (PSM H18), *Rhus trilobata* (PSM H16), and, at the top of the divide between La Joya and Lagua Zarca (2100 m.), *Arctostaphylos lanata* (PSM 136).

As in Cloud Forest, rock spires, rock castles, karst haycocks, and lapies produce topographic conditions unsuitable for the growth of forest (Pl. VI, Fig. 1). The combination of loose, sharp-edged limestone and equally sharp-spined agaves covering them makes many spires virtually unscalable. *Agave* (PSM H9), *Dasylirion*, and other plants typical of drier parts of the Sierra Madre occur here.

A shrub in disturbed sites, especially along trails, is the catclaw, *Mimosa* sp. (PSM H50). *Cnidoscolus*, a very common genus in disturbance areas in the Cloud Forest, is seldom found in the Pine-Oak area.

Pine-Oak Forest is found mainly on eastern slopes of the Sierra Madre front, covering virtually all areas above 1700 m. and interdigitating with Cloud Forest down to 1300 m. On the western (dry) side it may extend down to 1700 m., and below, in sheltered ravines. Gradually it is replaced by the following type.

Dry Oak-Pine Woodland. — Proceeding west from the top of the Sierra de Guatemala at 2100 m. along the La Joya trail, the woods begin to thin out, shrubs decrease in density, and grass appears in abundance. The oaks become round-crowned, their trunks are gnarled, and the trees may be either evergreen or deciduous. Within the Gómez Farías region above 900 m. this parklike woodland or savanna is usually dominated by oaks; however, pine or a mixture of pine and oak is found at La Joya de Salas and halfway between this locality and Carabanchel at La Joya Pinosa. The tallest trees in Oak-Pine Woodland seldom exceed 25 m. and most are under 20 m. in height. Epiphytic orchids, ferns, and Spanish moss are common on the oaks, but hepatics, true mosses, and other more delicate epiphytes are rare or absent, as are lianas. Sclerophylls typify most of the leaf types; compound-leaved trees are rare.

Some of the more common species include: *Quercus clivicola* (PSM 133), *Q. polymorpha* (PSM 163), *Q. grisea* (PSM H29), *Q. canbyi* (lower, drier slopes), *Pinus montezumae, P. teocote* (PSM H47), and *Juniperus flaccida* (PSM H23). Large agaves around La Joya are exploited for fiber and pulque. Madroño (*Arbutus*) is present in more humid woods.

The overgrazed La Joya basin is supplied with luxuriant pasture only in the summer rainy season. Cultural interference has probably been responsible for the spread of acacias through the basin at the expense of pine and oak. The nature of the original vegetation in the cultivated parts of the basin is a matter for speculation; I suspect it was largely pine on level areas and oak on hillsides.

The pine savanna at La Joya Pinosa is dominated by *Pinus teocote* (PSM 126), growing over chaparral, unlike that at La Joya de Salas, which is largely *P. montezumae* over grasses.

At Carabanchel the oak woods are fairly dense, and some shrubs typical of Humid Pine-Oak Forest are present. *Juglans pyriformis* (PSM 162) is mixed with various oaks.

Immediately north of the Gómez Farías region both west of Ciudad Victoria and west of Adelaida, *Quercus canbyi* (PSM 155), *Q. rysophylla* (PSM 034), *Q. vaseyana* (PSM 034), and *Q. clivicola* descend to low elevations (300-500 m.), where they adjoin Thorn Forest (Fig. 6). In the Gómez Farías region this altitude is occupied by Tropical Deciduous Forest or Tropical Evergreen Forest. The pine that appears at low elevation is usually *P. teocote* (PSM H46). West of Victoria, *Sabal texana* (= *S. mexicana*) is found in mixed Oak-Pine Woodland above 400 m., an interesting paradox to one accustomed to finding *Sabal* only in dense lowland palm bottoms.

Montane Chaparral (Pl. VII, Fig. 1). — Above 1700 m. on the flanks of rain shadow ridges very few trees are found, and grasses are replaced largely by low, dense thickets of scrub oak and other shrubs. The area occupied by this type is largely uninhabited, grazing is not heavy, and I believe this Montane Chaparral zone reflects a distinct climatic zone.

Very dense thickets of oaks may reach 1.8 m. in height and include a variety of other shrubs, scrub *Arbutus*, and scattered yuccas. The formation is evergreen, sclerophyll, medium to narrow-leaved, almost or exclusively shrubby, and not thorny. No attempt was made at a floristic inventory, and only four species were collected: *Quercus opaca* (PSM 131), *Q. grisea*, *Bauhinia coulteri* (PSM 127), and *Cercocarpus fothergilloides* (PSM 156).

Apparently this formation is quite similar to, if not actually identical with, Western Montane Chaparral described by Muller in Nuevo León (1939) and Coahuila (1947) where it also occurs on rain shadow slopes above 2000 m. Leopold included this formation in his Chaparral zone but did not attempt to map its distribution in eastern Mexico because of the small area involved.

At lower elevations, between 1300 and 1600 m. on the slopes above the Jaumave Valley, many shrubs grow; however, a scattering of low trees, mainly oaks, junipers, and palms (*Brahea*) 5 to 8 m. in height, and various grasses produce a savanna rather than chaparral structure. I have mapped these regions as Dry Oak-Pine Forest. Some shrubs encountered at 1500 m. southwest of Jaumave, which may or may not occur in typical oak chaparral, include *Dodonaea viscosa* (PSM 159), *Pistacia mexicana* (PSM 158), *Krameria cytisoides* (PSM 157), *Quercus grisea*, and *Dioon* sp.

Plant Indicators

The following 15 species were selected for their ease of identification and abundance in the field. If lumbering and other destructive interference have not proceeded too far, the more common trees and shrubs of a region are much easier to use as indicators of zonal change than are the less conspicuous, less abundant vertebrate animals. For example, in the Gómez Farías region fir and yew (*Abies* and *Taxus*) are sufficient to predict the presence of a cool, humid forest fauna with *Geophis semiannulatus, Rhadinea crassa,* and various salamanders.

Pinus spp. — The three pines of the Gómez Farías region increase in xeric tolerance in the following order: *P. patula, P. montezumae,* and *P. teocote. P. patula,* a "triste" or drooping-needle pine is found only in Humid Pine-Oak Forest above 1200 m. on the east side of the Sierra de Guatemala and above 1700 m. on the west (dry) side. A few tall individuals grow in Cloud Forest as low as 1200 m. *P. montezumae* is common above 1200 m. with a few individuals entering Cloud Forest at that elevation. Six which grew at the Rancho del Cielo clearing when Frank Harrison first arrived in 1935 may have been planted earlier, and the same may be true for those few individuals at Ejido Alta Cima, 850 m. West of the Humid Pine-Oak Forest *P. montezumae* occupies drier areas including the basin of La Joya de Salas and the Sierra de Tula. *Pinus teocote* is much less common than either of the former species, and in the Gómez Farías region occurs with *P. montezumae* from the top of the Sierra Madre to the La Joya basin. An isolated open grove in the Chaparral zone midway between La Joya de Salas and Carabanchel, called La Joya Pinosa, is exclusively *P. teocote.* Outside the Gómez Farías region *P. teocote* descends to 450 m. in dry oak-pine-*Sabal* woods near Ciudad Victoria. Two other pines, found in humid forest at high elevations west of Ciudad Victoria, *P. ayacahuite* and *P. pseudostrobus,* were not encountered in the Gómez Farías region.

Podocarpus reichei. — This conifer with its broad, willow-size leaves, is a good Cloud Forest indicator. It is found only between 950 and 1700 m. on the east side of the Sierra Madre. Above 1200 m. its local distribution is controlled by the development of dolines and other basins; it does not grow on dry ridges within the Cloud Forest.

Abies sp., *Taxus globosa.* — These two trees are generally found together, from as low as 1300 m. in upper Cloud Forest to the top of the mountain, and on the west side down to 1800 m. Unlike *Pinus patula,* which has a similar distribution, they are confined to more humid ravines and level valleys where soil has accumulated and drainage is not too rapid.

Arbutus ?xalapensis. — Near Rancho del Cielo the madroño is not common below 1600 m. and reaches its lower limit in pine-oak woods at about 1400 m. It is characteristic of Humid and Subhumid Pine-Oak Forest above 1600 m. A dwarf form grows in the Montane Chaparral zone.

Liquidambar styraciflua. — Sweet gum tolerates somewhat drier conditions than the more typical Tamaulipan Cloud Forest indicators such as *Podocarpus* and will grow as a low tree or one of medium height (8 to 10 m.) in areas approaching Montane Low Forest in aridity. There is one such area at Aserradero del Refugio (1) and another along the Antiguo Morelos – Ciudad Maíz highway across the Sierra Madre in San Luis Potosí. In the Gómez Farías region *Liquidambar* is found between 850 m. and 1700 m.; above 1300 m. it grows in ravines or on level ground, avoiding slopes. Areas of greatest abundance occur between 1000 and 1300 m.

Quercus germana. — This oak with entire-margined leaves ranges between 450 m. and 1200 m. through both Tropical Semi-Evergreen Forest and Cloud Forest. It is a good indicator of mild, rather humid areas at low to moderate elevation.

Q. sartorii. — A spinose-leaved oak (treated under the name *Q. skinneri*? by Harrell) which ranges between 900 and 1600 m. is a more reliable indicator of Cloud Forest than *Liquidambar* and is abundant in this habitat. Like other Cloud Forest species it is local above 1200 m.

Beaucarnea inermis. — This distinctive tree with a bulbous trunk and plumelike clusters of leaves grows on bare limestone rock between 300 and 600 m. in areas of Tropical Deciduous Forest. On the west side of the Sierra de Tamaulipas at 240 m. one was found in an open, cultivated field surrounded by organ pipe cactus, ebano, and huisache. All others were on limestone rock in the Tropical Deciduous Forest zone.

?*Dracaena* or ?*Beaucarnea.* — A tree 3 m. high resembling a yucca but with relaxed, drooping, rather than erect leaves is typical of Pine-Oak Forest above 1600 m.

Chamaedorea sp. and other small palms. — The very slender, sparsely leaved *Chamaedorea* appears in Cloud Forest (where not heavily shaded) and Tropical Evergreen or Semi-Evergreen Forest. Its local range also includes a few localities in Humid Pine-Oak forest above 1600 m. The lower altitudinal limit is in Semi-Evergreen Forest at 350 m. *Brahea*, another small palm with large fronds, barely enters Cloud Forest. It is a common shrub in Pine-Oak Forest. As a tree, *Brahea* occurs at 1200-1500 m. in oak woodland above Jaumave.

Cook (1909:13) found that the undergrowth palms, especially *Chamaedorea*, are absent in many localities affording apparently suitable conditions for their growth. "The undergrowth palms remain abundant only in regions which have not been completely deforested for agricultural purposes, and especially in districts too mountainous and broken for agricultural use." If such an indicator value can be ascribed to *Chamaedorea*, it supports my view that Tamaulipan Cloud Forest has, until recently, suffered little destruction by man.

Petrea arborea.— This blue-flowered vine is conspicuous in the dry season in Tropical Deciduous and Semi-Evergreen Forest from near sea level to 550 m. on the mountain slopes. West of Ocampo it reaches 900 m.

Guazuma ulmifolia. — Found only in Tropical Deciduous Forest, the "aquiche" tree reaches 700 m. on drier slopes north of El Tigre. It is abundant in the lowlands and apparently does not enter the Tropical Semi-Evergreen Forest zone.

Correlation of Vegetation Types

The zoological importance of recognizing homologies between habitats cannot be ignored. Before an effective integration of tropical biogeography and ecology is possible, this problem must be met. Which areas are equivalent as animal habitats? Is Short Tree Forest of Sonora an environmental homologue of Tropical Deciduous Forest in Tamaulipas?

TABLE V

Homologies of Mexican Vegetation Types

Gómez Farías Region, Tamps.	Mexico (Leopold, 1951)	Chiapas (Miranda, 1952)	Nuevo León (Muller, 1939)	Coahuila (Muller, 1947)	Río Mayo Basin, Sonora (Gentry, 1942)	Cerro Tancítaro, Michoacán (Leavenworth, 1946)
Thorn Forest and Scrub	Mesquite-Grassland (part), Thorn Forest	Sabana (part)	Eastern coastal plain Scrub	Tamaulipan Thorn Shrub	Thorn Forest	Open Arid Scrub Forest
Thorn Desert	Arid Tropical Scrub, Desert (part)		Central Plateau desert Scrub	Chihuahuan Desert Shrub		
Tropical Deciduous Forest	Tropical Deciduous Forest	Selva Baja Decidua			Short Tree Forest	Tropical Deciduous Forest, Dense Arid Scrub Forest
Tropical Semi-Evergreen and Evergreen Forest	Tropical Evergreen Forest	Selva Alta Subdecidua				
Cloud Forest	Cloud Forest	Bosques Deciduos (con Liquidambar), Selva Baja Siempre Verde				
Humid Pine-Oak Forest	Pine-Oak Forest	Encinares y Pinares	Montane Mesic Forest	Montane Mesic Forest	High Pine Forest	Upper Plateau Pine Forest
Dry Oak-Pine Woodland	Pine-Oak Forest	Encinares y Pinares	Montane Low Forest	Montane Low Forest	Low Pine Forest, Oak Forest	?Subtropical Transition Forest
Montane Chaparral	Chaparral		Western Montane Chaparral	Montane Chaparral		

The chief difficulty in measuring equivalence is the lack of an adequate, uniform, quantitative basis of comparison. Vegetation study has long been reduced to mere floristic inventory by many ecologists as well as by tropical plant taxonomists. The excessive use of this method is illustrated by the long lists of animals and plants, which Goldman (1951) offered in characterizing life zones in Mexico. The life zone concept emphasizes the value of recognizing climatic and biotic equivalence between areas; however, measurement of this equivalence is ineffective in terms of species lists alone.

While a taxonomic inventory may be useful for comparing adjacent or historically related areas, it diminishes in value with increasing distance. Short Tree Forest in Sonora (Gentry, 1942) is quite distinct in species composition from Tropical Deciduous Forest in Tamaulipas. Comparison between these areas gains value if it includes data on structure, function, and development of the vegetation plus climatic and other physical measurement.

While it begs the question, the following attempt at a correlation (Table V) indicates my interpretation of literature descriptions.

I have not listed Mexican vegetation types which do not occur in the Gómez Farías region. Some of these include Rainforest, Tropical Savanna, Boreal (Subalpine) Forest, and Alpine Meadows. I have attempted to correlate only climax types, excluding edaphic variants as pedregal, gallery forest, and so forth.

The following two comments on Cloud Forest appear to be pertinent: Two vegetation types each of which might be considered Cloud Forest exist in southern Mexico (Harrell, pers. comm.). In the mountains of Chiapas

Miranda's "Bosques Deciduos (con Liquidambar)" and his "Selvas Bajas Siempre Verdes" are both comparable to Tamaulipan Cloud Forest in certain features. The name Deciduous Forest which Miranda (1952) uses may be misleading; a translation of his description (p. 137) clearly implies a semi-evergreen forest: "A certain number of the most characteristic trees lose their leaves completely or almost completely during certain winter months from November or December to February. However, this phenomenon does not have the universality displayed in the north temperate zones since a high proportion of trees that do not lose their foliage or that renew it rapidly always exist in the deciduous forest [of Chiapas]."

Valley Forest between 7000 and 9000 feet in Michoacán (Leavenworth, 1946), subdivided into Transition to Cloud Forest and Cloud Forest, is not equivalent to this formation in eastern Mexico. The heavy epiphytic growth of ferns, mosses, liverworts, and lichens that Leavenworth took as an indication of Cloud Forest conditions occurs elsewhere throughout the Plateau in boreal or humid subalpine forest. These cool forests have a lower mean annual temperature and a greater frost frequency than that tolerated by typical broad-leaved Cloud Forest.

Summary

Eight formations recognizable in terms of structure, function, and flora are found in the Gómez Farías region. They appear equivalent to eight of the 12 Vegetation Zones recognized by Leopold in Mexico. Presumably these appear in response to important differences in climate. Three, Cloud Forest, Tropical Deciduous Forest, and Tropical Semi-Evergreen Forest, are unknown north of the Gómez Farías region. All montane types, Cloud Forest, Tropical Semi-Evergreen Forest, Pine-Oak Forest, Oak-Pine Woods, and Montane Chaparral, are isolated on the Sierra de Guatemala, and surrounded by arid lowland formations.

A close correspondence appears to exist between five of the types found in the Gómez Farías region and the vegetation described by Muller in Nuevo León and Coahuila. Vegetation types in other parts of Mexico are less closely equivalent.

AMPHIBIANS AND REPTILES OF THE GOMEZ FARIAS REGION

Aside from aquatic forms the herpetological fauna of the Gómez Farías region comprises 94 species. I have attempted to characterize the local distribution of each in terms of altitude and vegetation type. Under each species an initial paragraph lists localities, elevations, and number of specimens collected. This is followed by information on habitat preferences and any noteworthy data on life history or food habits. When such a discussion appears pertinent, I have commented on the ecological behavior and distribution of the species outside the Gómez Farías region.

With regard to systematics I have outlined briefly only some of the current problems. My purpose is to describe local ecology and distribution rather than to treat subspecies in detail, or to describe population parameters for all species in this area. Such data may make future studies more meaningful in ecological as well as in taxonomic terms. In general, I have kept the taxonomic treatment to a minimum. This does not imply that serious taxonomic problems do not exist.

For purposes of zonal analysis in a transect study trinomial determinations are not of primary interest. Only *Masticophis taeniatus* and possibly *Rana pipiens* and *Sceloporus variabilis* are represented by more than one subspecies in the Gómez Farías region. For purposes of ecological and evolutionary studies through the range of a species, knowledge of taxonomic units below the species level may be quite important. In several poorly known species, as *Rhadinaea crassa*, *Chiropterotriton chondrostega*, and *Abronia taeniata*, the collections from the Gómez Farías region will be of value in describing population variation.

Most of the specimens from the Gómez Farías region are in the collection of the Museum of Zoology of the University of Michigan (UMMZ). Usually these are cited by locality only, with number of specimens in parentheses. Specimens from other collections are indicated thus: TU — Tulane University Department of Zoology; MMNH — Minnesota Museum of Natural History; AMNH — American Museum of Natural History; USNM — United States National Museum, EHT — Edward H. Taylor.

All collecting stations listed are either given in Table I or are described in terms of distance from these localities. Other localities outside the Gómez Farías region are given on the American Geographical Society millionth maps or on the World Aeronautical Charts, U. S. Coast and Geodetic Survey. All distances and elevations are expressed in terms of the metric system. Most elevations were checked on more than one occasion and are considered accurate to within 50 m.

Only terrestrial reptiles and amphibians in the Gómez Farías region are treated in this account. Thus the following largely aquatic species are excluded: *Kinosternon integrum*, *K. cruentatum*, *Pseudemys scripta*, and *Crocodylus moreleti*. With few modifications nomenclature and systematic presentation is that of Smith and Taylor (1945, 1948, 1950).

Order Caudata, Salamanders

Pseudoeurycea belli. —Rancho del Cielo, 1050 to 1200 m. (6); 2 km. NW Rancho del Cielo, 1400 m. (2); near Lagua Zarca, 1600 m. (2); Agua Linda, 1800 m., PSM Field No. 3028 (specimen lost); total, 11 specimens.

Of the five salamanders in the Gómez Farías region, this is the largest and the least abundant. Except for two juveniles found together below a stone, all individuals collected were solitary. Near Rancho del Cielo a juvenile appeared 9 m. below ground level on the floor of a calcite cave called Crystal Cave by Frank Harrison; one other specimen, an adult from Agua Linda, was found in a cave. A large adult came from the bottom of a deep open sink, 25 m. below ground level near Rancho del Cielo. All others were taken on the ground under cover, usually wet logs. From the mountainside west of Rancho del Cielo one was dug out of red soil and loose rock by a road-constructing crew. The stomach of a gartersnake, *Thamnophis mendax*, contained remains of another. Walker and Heed recovered several hatchlings under wet logs in late August.

Despite its abundance through the pine-oak belt of the Sierra Madre Oriental and the Transverse Volcanic Province of the Central Plateau, *P. belli* evidently does not follow pine-oak forests northward in the Sierra Madre Occidental beyond Nayarit. Its discovery at 23° N lat. in Tamaulipas, 180 km. north of the nearest previously known locality (Xilitla, San Luis Potosí, Taylor, 1949), was not anticipated. In Tamaulipas it inhabits Cloud Forest and Humid Pine-Oak Forest between 1050 and 1800 m.

In several features Tamaulipan specimens differ from typical *belli* of the Transverse Volcanic Province. In the adults the paired red spots of the dorsum tend to be fused, forming red chevrons; this may be an ontogenetic development since it is not apparent in five juvenile specimens. The red saddle above the forelegs extends down on the sides almost to the level of the arm insertion; in typical *belli* it is confined to the dorsum. Finally the adpressed limbs of two males are separated by only one costal groove, rather than by three as in typical *belli.* Two individuals from Guerrero, Hidalgo, the locality nearest to Tamaulipas from which I have seen material, do not differ appreciably from Michoacán specimens in any of the above features. Critical examination of a larger series may justify taxonomic recognition of the Tamaulipan population.

Pseudoeurycea scandens. — Rancho del Cielo, 1000-1250 m. (62); Rancho Viejo, 1200 m. (16); Valle de la Gruta 3 km. W of Rancho del Cielo, 1500 m. (9); Aserradero del Inferno, 1 km. S of La Gloria, 1340 m.; trail to Lagua Zarca, 1350 m. (7); Lagua Zarca, 1600 m. (3); Agua de los Indios and vicinity, 4 km. SSW of Rancho del Cielo, 1200-1300 m. (16); Agua Linda, 1800 m. (3); total, 116 specimens, all in Cloud Forest or Humid Pine-Oak Forest.

With the addition of two specimens from a cave at Chihue, 1850 m., northwest of Ciudad Victoria, the above localities represent the known range of this recently described species (Walker, 1955*b*).

Lacking any effective comparative sampling technique I cannot judge whether *P. scandens* is more abundant in subterranean habitats or below cover at the surface. Sixty-one specimens came from caves, whereas 47 were taken in terrestrial situations, within and under logs, rocks, or moss. Certainly the karst caves and sinkholes inspected in the Rancho del Cielo region represent a very small percentage of the total available subterranean environment. The deepest caves explored were more than 30 m. in depth, but no salamanders appeared below 15 m. and most were much closer to the surface. As noted by Walker (*l.c.*), cave walls and ceilings are favorite retreats; I cannot recall finding any on the floor. The largest aggregation was found in a small, dimly lighted fissure 3 m. below ground at about 1500 m. in Humid Pine-Oak Forest near Casa Piedras. Fourteen *P. scandens* and one *Chiropterotriton multidentata* were scattered about the narrow, dripping walls in small groups. One cave cricket was the only evident arthropod. Of roughly 20 caves and sinks explored at Rancho del Cielo less than half harbored salamanders.

Although less numerous in bromeliads than either species of *Chiropterotriton,* five specimens of *P. scandens* were taken in these plants. In a total of 41 bromeliads searched at Rancho Viejo in July, 1950, I found 11 *Chiropterotriton* and two *P. scandens,* the latter at 2.1 and 2.5 m. above the ground. Two were found under bark of standing, rotting saplings.

The lower altitudinal limit of *P. scandens* is not known precisely but may correspond to that of the Cloud Forest. Significantly, *scandens* did not appear in the wet cave at Aserradero Paraíso (420 m.) where its amphibian associates in the Rancho del Cielo caves, *Eleutherodactylus hidalgoensis*, *Syrrhophus latodactylus*, and *Chiropterotriton multidentata*, all were found.

For discussion of relationships, morphology, and further ecological notes see Walker (*l.c.*).

Pseudoeurycea cephalica. — Rancho del Cielo and vicinity, 1000-1200 m. (19); trail to Agua de los Indios 3 km. SW of Rancho del Cielo, 1300 m.; Agua Linda, 1800 m.; total, 21 specimens from Cloud Forest and Humid Pine-Oak Forest.

In addition to those found below cover several were inside logs in the tunnels of beetle larvae (*Passalus?*). Although perhaps less often found inside logs, *P. scandens* occupies the same terrestrial habitats as *cephalica* and is definitely the more numerous. Unlike *scandens*, *cephalica* has not been found in caves or bromeliads.

P. cephalica is unknown north of the humid montane forests of western Tamaulipas. North of the Gómez Farías region it occurs at 1500 m., east of Dulces Nombres. As with other elements of the "northeast Madrean" component, a large distributional hiatus through San Luis Potosí separates northern and southern populations of the species (Table 8). The Rancho del Cielo specimens have been identified as *Pseudoeurycea cephalica rubrimembris* Taylor and Smith (Walker, 1955*b*).

Chiropterotriton multidentata. — Cloud Forest, Humid Pine-Oak Forest, and Tropical Evergreen Forest at Aserradero del Paraíso, 420 m. (89); Rancho del Cielo and vicinity, 1000-1200 m. (65); North Woods 2 km. E of Lagua Zarca, 1350 m. (2); Rancho Viejo, 1200 m. (7); Valle de la Gruta 3 km. WNW of Rancho del Cielo, 1530 m. (2); Agua de los Indios, 4 km. SSW of Rancho del Cielo, (4); San José, *ca.* 2 km. NW, 1530 m.; Aserradero Refugio (No. 2), W of La Gloria, 1680 m.; Agua de los Perros, 1 km. N of Agua Linda, 1860 m. (4); 2 km. S of Agua Linda, 1860 m.

Of the 176 identifications listed above, habitat data are available on 174 as follows: 103 were collected in caves, two from under ground cover, and 69 from bromeliads or rotting tree branches. Caves in the Cloud Forest and Pine-Oak Forest accounted for only 14 specimens, most of these representing one or two individuals per cave. The rest, 89, came from three separate visits to a very damp cave 1 km. S of Aserradero del Paraíso in Tropical Evergreen Forest, elevation 420 m. The remarkable aggregation of salamanders at Paraíso far exceeded that of any other Tamaulipan caves I have investigated. A feature contributing to the wealth of fauna may be a permanent pool of water, fairly close to the mouth of the cave. This is responsible for unusually cool, humid air, which extends several meters outside the cave mouth. Neighboring caves do not produce the moisture drip evident in this cave, although their relative humidity is probably close to saturation.

Amphibians were found only near the mouth of the cave, some in fairly bright light. None appeared in the deeper recesses 10 m. below the surface. Smooth sides of the walls as well as ledges, cracks, and, in a few places, the cave floor, were occupied.

C. multidentata is more common in bromeliads than any of the other Tamaulipan plethodontids. As many as five were discovered in one big tank bromeliad which had a rich invertebrate fauna, including centipedes, crickets, spiders, snails, slugs, and mosquito and other insect larvae. The plants investigated were 0.5 to 5 m. above ground. The remarkable freeze of early February, 1951, virtually extirpated tank bromeliads in the Rancho del Cielo Cloud Forest and may have decimated the arboreal salamander fauna. Strangely, bromeliads above 1200 m. were not injured and continued to yield salamanders in 1951 and 1953.

Although I have no comparable data, I suspect that bromeliads are slightly more productive in the dry season. Thirty-seven salamanders were collected in one day in bromeliads near Rancho del Cielo on May 5, 1949, a total not equaled by Walker and Heed in August and September of 1950. The most they obtained in a single day was four, from over a dozen plants at Agua de los Indios. Of a total of over 50 bromeliads torn open at Rancho Viejo in late July and early August of 1950 only six specimens of *C. multidentata* were obtained. In that area *C. chondrostega* may have been slightly more numerous; ten of the latter were collected in bromeliads.

In addition to bromeliads, rotten snags or limbs are occupied; in one such snag four were taken from under the bark in ant tunnels 1.8 to 3.0 m. above ground.

A member of the "northeast Madrean" component, *C. multidentata* is known from at least four localities in Humid Pine-Oak Forests south of Tamaulipas. North of the Rancho del Cielo area I found it at Chihue, west of Victoria, in caves between 1860 and 2400 m.

It is not always possible to differentiate *C. multidentata* and *C. chondrostega* in the field. The former species attains a larger size than *C. chondrostega*, and all individuals larger than 65 mm. in total length can be assumed with reasonable certainty to represent the former. Adults of *multidentata* usually exceed 80 mm. in total length, whereas *chondrostega* adults range between 50 and 62 mm.

I am indebted to George B. Rabb for identifications of all Tamaulipan *Chiropterotriton.*

Chiropterotriton chondrostega. — Rancho del Cielo, 1000 to 1150 m. (65); Agua de los Indios and vicinity, 4 km. SSW of Rancho del Cielo, 1200-1300 m. (7); Rancho Viejo, 1200 m. (11); above Casa Piedras, 1450 m.; North Woods and trail below Lagua Zarca, 1350-1450 m. (14); Lagua Zarca, 1600 m.; Agua Linda, 1800 m. (46); La Lagunita, *ca.* 4 km. by road WNW of La Gloria, 1890 m. (3); total, 148 specimens, all collected in Cloud Forest or Humid Pine-Oak Forest.

This species bears roughly the same ecological and morphological relationship to *C. multidentata* that *P. cephalica* holds to *P. scandens.* It is smaller than *multidentata* and has relatively shorter legs; it is not found in caves, and although it will climb, it is less numerous in bromeliads than *C. multidentata.* Thirty-two individuals came from bromeliads in contrast to 69 of *multidentata,* a ratio between the species of about 1:2. As no attempt was made to identify the two in the field, there should be no collecting bias. I am not sure if the two occur together in the same bromeliads; however, on several occasions at both Rancho del Cielo and Rancho Viejo both species were collected the same day from bromeliads in one small area.

Although not found inside caves, *C. chondrostega* does occur on the bottom of open sinks to a depth of 25 m. below the surface of the ground. North of Rancho del Cielo, a large bell-shaped sink of this type with overhung sides contained six *C. chondrostega* and one *P. belli* under an accumulation of duff, debris, and rotten branches in the bottom. Two other sinks, 12 m. in depth with some vegetation growing at the bottom, each yielded eight *C. chondrostega.*

In more conventional terrestrial habitats *C. chondrostega* was collected in or under rotten logs. On March 18 at Agua Linda I systematically stripped loose plates of bark for a distance of 30 m. along a large pine log 1.3 m. in diameter, finding 45 salamanders of this species, one *P. cephalica,* and two *Rhadinaea.* A smaller section of a large rotten sweet gum in the North Woods area north of Rancho del Cielo yielded twelve. Aggregations of this size are uncommon; many other logs were searched without success. Presumably the rotten logs, rejected railroad ties, and other waste lumber left after recent timber exploitation will provide a tremendous, if temporary, expansion of terrestrial salamander habitat.

Outside the Gómez Farías region *C. chondrostega* is known only from the type locality, Durango, Hidalgo. Neither Walker, Harrell, nor I found it in our brief visits to oak-sweet gum and pine forests northwest of Ciudad Victoria. As noted under *C. multidentata,* identification of Tamaulipan *Chiropterotriton* may be difficult, and specimens of small size cannot be distinguished satisfactorily under most circumstances. Sexual dimorphism obscures the specific differences in tooth number and leg length. Adpressed limbs of *chondrostega* females are separated by three costal grooves, of *multidentata* by one or less; in males the toes of the former are separated by one groove, they touch or overlap in the latter.

Order Salientia, Frogs

Rhinophrynus dorsalis. — Arid as well as humid tropical lowlands in northeastern Middle America characterize the known range of this unique species. I am aware of only three Tamaulipan locality records, all in the arid lowlands. In northern Tamaulipas at San Fernando after a torrential rain, *Rhinophrynus* was discovered in June, 1951 (Davis, 1953). At Hacienda La Clementina near Llera just outside the Gómez Farías region, Bryce Brown found breeding individuals. The two specimens from the Gómez Farías region are immature. These were collected and preserved by Dan Cameron 2 km. east of Chamal, 200 m. elevation. Mr. Cameron informed me that he discovered them during his fall plowing in 1950 and 1951.

Scaphiopus couchi. — San Gerardo, 4 km. W, 120 m. (3); Villa Juarez (=Ciudad Mante), *ca.* 80 m.; total, 4 specimens.

The spadefoot is a member of the "arid interior" faunal group. Its range does not extend south into heavy forest.

Bufo horribilis. — Chamal, 6 km. NE, 200 m. (8); Pano Ayuctle, 100 m. (10); Gómez Farías, 360 m. (2); Rancho del Cielo and vicinity, 900-1150 m. (7); total, 27 specimens.

Clearings in the Cloud Forest at Rancho del Cielo probably mark the upper altitudinal limit of this species. In the Tamaulipan lowlands it is common, but usually confined to the vicinity of wells or ponds, especially in the dry season.

Bufo valliceps. — La Union, 120 m., TU 15810; Pano Ayuctle, 100 m. (14), TU 15810; Gómez Farías, 360 m. (11); Aserradero del Paraíso, 420 m.; trail between Pano Ayuctle and Rancho del Cielo, *ca.* 900 m.; Rancho del Cielo, 900-1200 m. (3); near Aserradero del Refugio No. 1, N of El Tigre, 1050 m. (6); La Joya de Salas, 1520 m. (4); total, 42 specimens.

Breeding may begin in early spring; in late March, 1949, strings of toad eggs, probably of this species, were found in a water hole northeast of Rancho del Cielo. Most reproduction, however, undoubtedly occurs during the rainy season when innumerable temporary ponds flood the lowlands and parts of the Sierra.

One individual was recovered from the stomach of a snake, *Leptodiera maculata,* collected near San Gerardo.

Among the toads of the Gómez Farías region *B. valliceps* is the most common and widespread. It is abundant throughout the range of *B. horribilis* in various arid and humid lowland forests, occurs in Cloud Forest, and is less numerous but present in certain parts of the Dry Oak-Pine Woodland. Individuals near La Joya were calling in small ponds in late July. The latter locality, 1550 m., is notably high for this lowland and foothill species. In the Alta Verapaz of Guatemala, Stuart (1948) determined its upper altitudinal limit at 1250 m.

Bufo punctatus. — A member of the "arid interior" faunal component, which includes such xerophiles as *Scaphiopus, Holbrookia,* and, among the mammals, *Citellus* and *Perognathus,* this toad has been recorded in the Gómez Farías region only from the dry Jaumave Valley, 750 m. To my knowledge *punctatus* does not follow arid lowland Thorn Scrub south of the Sierra de San Carlos region (Gaige, 1937). A wide zone of increasing precipitation in southern Tamaulipas and northern Veracruz acts as a gradual climatic barrier to this and the other lowland xerophiles.

Just west of the Gómez Farías region *B. punctatus* is common in the vicinity of Tula and Palmillas; it ranges much farther south on the Central Plateau than on the coastal plain, following arid basins and rain shadow slopes south at least to Guanajuato.

Syrrhophus smithi. — La Joya de Salas, 1530 m. (4). On dry rocky slopes of the oak-pine woods about La Joya de Salas a *Syrrhophus* is common in the rainy season, to judge from calls; however, like all *Syrrhophus* that do not aggregate in caves, it is not easy to collect. Only four specimens, all males, were obtained on the night of July 21 when their feeble calls, both single notes and short trills, were heard scattered over a hillside (Lidicker and Mackiewicz).

These specimens are referred to *S. smithi* with some hesitation; they are, however, quite different from *S. latodactylus* and *S. cystignathoides,* the other two species in the Gómez Farías region.

Syrrhophus latodactylus. — Aserradero del Paraíso, 420 m. (18); San Pedro mine 3-4 km. WNW of El Carrizo, 560-580 m. (33); 5 km. N of Aserradero del Paraíso, 750 m. (3); vicinity of Rancho del Cielo, 1050-1200 m. (39); lower edge of Pine-Oak Forest, 2 km. E of Lagua Zarca,

1350 m.; between the latter and Lagua Zarca, 1540 m.; Agua Linda, 1800 m. (3); total, 98 specimens. Adults were found in caves, mine shafts, and under cover on the ground in Tropical Deciduous Forest, Tropical Evergreen Forest, Cloud Forest, and part of the Humid Pine-Oak Forest. From two abandoned shafts at San Pedro mine, Harrell and I captured 33 specimens in a few hours; other localities were less productive. Over ten visits to a calcite cave (Crystal Cave) near Rancho del Cielo yielded 17 specimens in a five-year period, with none present on many days and no more than five found at one time.

Outside the Gómez Farías region I have collected this species in a cave at El Chihue northwest of Ciudad Victoria, elevation 1900 m. A series of 26 from Cueva El Pachon near Antiguo Morelos (250 m.) may also represent *S. latodactylus*, although they differ from Gómez Farías specimens in several respects.

Damp caves and mine shafts in regions of karst limestone are favorite retreats through the dry season and may continue to be inhabited in the rainy season although more are found aboveground at that time. Most of the specimens were obtained in troglodytic habitats which *S. latodactylus* shares with *Eleutherodactylus hidalgoensis*, *E. latrans* (low elevations only), and various other salamanders. Smaller individuals, those under 20 mm., are seldom found in caves; they usually appear under rocks or debris on the ground, and are associated with *S. cystignathoides*. During rains and damp weather adult males may be heard calling aboveground between rocks or under cover. Presumably they do not breed in caves.

The call is a soft, insect-like "kip" or "chip," easily imitated by a short, forced whistle. An imitation will often induce the frogs to start calling, and in this manner specimens were located in several large caves that at first appeared uninhabited; a large collection was made at San Pedro, using this method. No breeding activity has been observed although ovaries of most of the large females collected in late winter and spring contained enlarged ova 2 mm. in diameter.

S. latodactylus ranges from northern Nuevo León south along the eastern escarpment and foothills of the Sierra Madre Oriental to Jacala, in northern Hidalgo. The main taxonomic problem centers on the relationship of this species to the other large, wide-disced members of the genus, *S. marnocki* and *S. gaigeae*, a relationship Taylor (1940) did not discuss in his description of *latodactylus*. Milstead *et al.* (1950) have subsequently synonymized *gaigeae* with *marnocki*. On the basis of specimens from the Gómez Farías region I find no grounds for synonymizing *latodactylus* and *marnocki*. A dark interorbital bar, longer legs, wider toe discs, and absence of male vocal slits are features of *latodactylus* in this area. It should be stressed, nevertheless, that a considerable gap separates Big Bend populations of *marnocki* in Texas from *latodactylus* in Nuevo León, and a contiguous allopatric distribution may be anticipated.

The series from El Pachon is distinct from Gómez Farías populations in having well-developed vocal slits in males and a vermiculate rather than a blotched pattern.

Syrrhophus cystignathoides.— Rancho Pano Ayuctle, 100 m. (7); La Union, 130 m., TU 15508, TU 15580 (9); 3 km. NW of Rancho Pano

Ayuctle, 500 m.; Aserradero del Paraíso and vicinity, 500-800 m. (14); Rancho del Cielo and vicinity, 1000-1200 m. (15); total, 46 specimens. All individuals were collected on or near the ground in damp places under logs, rocks, chips, and other debris; none appeared in caves. Young individuals of *S. latodactylus* occur in similar habitat above 300 m., and without close examination the two may be confused in the field. *S. cystignathoides* is usually pinkish, salmon, or yellow-tan dorsally, especially on the femur; *latodactylus* varies from tan to olive green with gray-brown or greenish yellow on the femur. *S. latodactylus* has relatively longer legs, a larger tympanum, and wider toe discs than has *cystignathoides*. *S. latodactylus* does not range over the Tamaulipan coastal plain and into the Sierra de Tamaulipas as does *S. cystignathoides*. Thus the two are sympatric only in the Sierra Madre Oriental.

In August in the rainy season Walker and Heed found this frog common at Pano Ayuctle, calling at night and occasionally in the day. In field notes they described the call as five to seven notes accelerating toward the end, "twit-twit-tit tit tit tit"; occasionally they heard a single trilled note. The night of August 8 they found eight individuals, six of these on the upper surface of leaves a meter or less above the ground. Many were heard in shrubs and dense cover in gallery forest along the Río Sabinas, less commonly along paths at the edge of cultivated fields and once in a sugar-cane field. At Aserradero del Paraíso in February I heard no calls, but in half a day collected 15 specimens under cover, especially under piles of chips, in partly cleared Tropical Evergreen Forest. At Rancho del Cielo, Heed and Walker found them singing nightly but not persistently, several weak, rapid chirping notes, not recognizably distinct from the calls heard at Rancho Pano Ayuctle earlier that month. In 15 days of diligent collecting they obtained 14 specimens including one in a flowerpot in Harrison's dooryard.

Tropical Evergreen Forest and Tropical Deciduous Forest are the centers of abundance of *S. cystignathoides* in southern Tamaulipas although two specimens (88242) come from either Thorn Forest or Gallery Forest along the Río Guayalejo near Magiscatzin. The upper altitudinal limit is reached in the lower Cloud Forest.

S. cystignathoides is quite similar to *S. campi* of south Texas and northeastern Mexico. I am unable to make an absolute distinction between them although the granular abdomen, smaller size, relatively wider toe tips, and reddish or tawny color serve to identify a large percentage of Tamaulipan *cystignathoides*. Specimens from the Sierra de Tamaulipas resemble Texan *S. campi* in size and ventral skin texture, but are close to *cystignathoides* of the Gómez Farías region in disc width and body color.

Leptodactylus labialis. — Pano Ayuctle, 100 m. (9); 4 km. W of San Gerardo *ca.* 120 m. (5); total, 14 specimens. During the rainy season in May, June, July, and August the soft, distinctive calls of this frog may be heard nightly during or after heavy rains. Small temporary pools, poorly drained hollows, or even open cultivated fields without standing water are occupied by singing males. The call is a short, whistled "whoot" or "wheat," readily imitated, and uttered about once a second when the frog is calling actively. As is typical of the other Tamaulipan leptodactylids, the males do not aggregate, and the problem of locating a single individual under cover of grass or, more usually, in a shallow burrow, can require much time and effort. Near San Gerardo I dug one out of a tiny hollow in a furrow of a plowed field. A few have been collected moving aboveground at night; these were not calling.

Although none was heard outside of the Río Sabinas lowlands, I would expect to find them in the cleared valleys of Chamal and Ocampo. Foothills of the Sierra Madre Oriental and rocky areas generally are avoided.

L. labialis ranges widely through the tropical lowlands of Middle America from the lower Río Grande Valley of Texas to Panamá.

Leptodactylus melanonotus. — From Pano Ayuctle, 100 m., 22 specimens comprise the only definite records although individuals were heard in cattail-lined irrigation ditches near Limón and at the Xicotencatl road junction north of Limón. Unlike *L. labialis* this species frequents permanent water, calling and probably breeding along streams and canals during the dry season. I have never heard it in the fields and temporary wet spots inhabited by *L. labialis*; presumably the two are ecologically isolated. The song is unmistakably different from that of *L. labialis*, a faint single "tuck" similar to the sound made when two small stones are struck together, uttered under hanging roots, rotting leaves, and grass at the edge of a stream or pool. Although suitable calling places are more restricted than in *L. labialis*, singing males are just as difficult to detect. At Pano Ayuctle most individuals were found near the drowned mouth of a small tributary, the Arroyo Encino, rather than along the Río Sabinas itself.

L. melanonotus ranges throughout the tropical lowlands of Middle America, reaching its northern limit in southern Tamaulipas. I am unaware of records from north of the Gómez Farías region, although the species occurs near Aldama and probably throughout southern Tamaulipas south of the Tropic of Cancer.

Eleutherodactylus hidalgoensis. — From the Gómez Farías region 64 of 68 specimens were taken in Cloud Forest near Rancho del Cielo between 1050 and 1200 m. (61); TU 15518; TU 15530; MMNH (5-6-54). Four others close to the Rancho del Cielo area came from the following habitats: Aserradero del Paraíso, 420 m.; Tropical Evergreen Forest, (2); Aserradero del Inferno, 1 km. S of La Gloria, 1320 m.; Upper Cloud Forest; near La Joya de Salas, *ca.* 1600 m., Pine-Oak Woodland. One other Tamaulipan record, outside the Gómez Farías region, was obtained at Chihue, 1830 m., in a cave in Humid Pine-Oak Forest. Of a total of 69 Tamaulipan specimens all but four were collected in caves or sinks; one was found about 30 m. below ground in a deep sinkhole, but most were taken closer to the surface. Niches and crannies well above the cave floor were favorite retreats. One individual was encountered just outside the dark mouth of a small cave; the rest were in total darkness or very dim light. Food contents of several stomachs examined by Dr. T. Hubbell included cave crickets *(Amphiacusta)*, a stone cricket *(Stenopalmathus)*, phalangids, a snout beetle, and a Hemipteran (?Cydnidae), all to be expected around, or a short distance within, cave mouths.

Two of three specimens collected aboveground were calling at night from the top of moss-covered boulders. None was found, or heard, in trees as Taylor reported (1942*b*). In field notes the call is described as either a resonant chub-chub-chug, an ak-ak-ak-akak, or "garum-pet." Presumably late winter and spring rains initiate calling. Time of egg-laying is unknown.

While Cloud Forest appears the center of abundance of *E. hidalgoensis* in the Gómez Farías region, the species has a slightly wider range than this, descending in humid valleys to Aserradero del Paraíso and ranging up into subhumid Pine-Oak Woodland near La Joya. Outside the Gómez Farías region the species is virtually unknown, only two locality records, Tianguistengo, Hidalgo and Tequeyutepec, Veracruz, are listed by Smith and Taylor (1948), with a third added from Xilitla, San Luis Potosí (Taylor, 1949). Collecting from caves in mountain karst terrain elsewhere should yield additional specimens.

E. hidalgoensis belongs to the "northeast Madrean" faunal component (Map 3). *E. spatulatus, decoratus, alfredi,* and possibly *xucanebi* of Guatemala are close relatives, although critical morphological study might reveal convergence in the diagnostic feature of the *alfredi* group, e.g., the wide toe discs.

Apparently *E. hidalgoensis* is the largest member of the *alfredi* group. Adult females usually exceed 50 mm. snout-coccyx length, and one measured 55.2 mm. Males are considerably smaller; the largest I measured was 44.9 mm. Tympanum size is not strikingly dimorphic.

Eleutherodactylus augusti. — La Joya de Salas, 1530 m. (5); Sierra Gorda, 3 km. WNW of El Carrizo, 450 m.; Valle de Paraíso, 420 m.; total, 7 specimens. Elsewhere the distinctive call of this species was heard at the Seymour Taylor Ranch, 6 km. NE of Chamal, *ca.* 200 m.; near Aguacates, 2 km. NW Gómez Farías, 650 m.; near Aserradero del Paraíso, *ca.* 300 m.; between El Tigre and Aserradero del Refugio No. 1, 1210 m.; and at Carabanchel, May 1, 1953, 2000 m. In southern Tamaulipas outside the Gómez Farías region *E. augusti* is known from Cueva El Pachon near Antiguo Morelos, 200 m.; near Palmillas, 1440 m.; Sierra de Tamaulipas, *ca.* 1000 m.; and near Tula (calling in July, 1951), *ca.* 1000 m.

Strictly saxicolous, *E. augusti* occupies all available rock habitats in southern Tamaulipas except Cloud Forest and possibly part of the Humid Pine-Oak Forest. In view of its apparent abundance in the Gómez Farías region, the scarcity of other records in eastern Mexico is unusual. Smith and Taylor (1948) list only two localities. During the dry season a few individuals were found in caves (El Pachon and El Carrizo) with *Syrrhophus latodactylus.* The abundance of *augusti* is best appreciated when the males call vigorously after occasional spring rains or during the start of the summer rains. The scarcity of specimens reflects a solitary habitat and wariness rather than lack of numbers. In Valle de Paraíso an evening's collecting yielded a single male in an area where at least 20 others were calling from within a radius of 500 m. When seized, this individual urinated and inflated himself until rigid in the manner reported by Wright and Wright (1949).

The only apparently suitable habitats not definitely known to be occupied by this frog in the Gómez Farías region, Cloud Forest and Humid Pine-Oak Forest, are inhabited by a large, gray-green relative with greatly expanded finger tips, *E. hidalgoensis*, suggesting a zonal replacement. Although there is some altitudinal overlap, I have not found the two in identical habitats. *E. hidalgoensis* in a cave in Tropical Evergreen Forest at Aserradero del Paraiso, 420 m., was below the elevation of, and within a few kilometers of, adjacent Tropical Deciduous Forest inhabited by *augusti*. Although both occur near La Joya de Salas, possibly in the same habitat, *E. hidalgoensis* is known at that locality from only one specimen. No *augusti* are recorded from Cloud Forest at Rancho del Cielo, although several unidentified calls may well represent this species. *E. augusti* occurs at Aserradero del Refugio No. 1, 1050 m., an area of low, scattered *Liquidambar* and oaks, a drier habitat than that found at Rancho del Cielo.

In its local distribution in southwestern Tamaulipas, *E. augusti* is virtually unique, ranging through both the treeless, xeric, *Yucca*-covered hills of the Tula district and into various forest habitats. The latter include subhumid montane oak-pine woods, dry oak-sweet gum forest, and lowland Tropical Deciduous Forest. Richard Zweifel has identified these specimens as *E. augusti augusti*.

Phrynohyas spilomma. — Widespread through the humid tropical lowlands of Middle and South America, *spilomma* reaches its northern distributional limit in Tropical Deciduous Forest of the Gómez Farías region. Two specimens, both collected by R. Darnell at Pano Ayuctle, 100 m., were found as follows: 104096, May, 1950, in a cavity of a large nopal cactus; TU 15444, December 21, 1952, captured at night perched on the broad leaf of an elephant ear *(Xanthosoma)*. In the humid nights of early June, 1953, I heard both *Smilisca baudinii* and *Phrynohyas* calling infrequently from trees near Pano Ayuctle.

In a recent taxonomic revision of this species, formerly known as *Hyla venulosa*, Tamaulipan specimens are cited (Duellman, 1956).

Smilisca baudinii. — An abundant species in the lowlands during the rainy season, less numerous in the Sierra Madre as high as 1250 m. Pano Ayuctle, 100 m. (20); TU 15447; San Gerardo, 3 km. W, 120 m. (2); E of Chamal, 150 m. (2); Gómez Farías, 360 m. (6); slopes of Sierra Gorda, 3 km. WNW of El Carrizo, 450 m.; Rancho del Cielo, 1050-1200 m. (20); MMNH June, 1953, May, 1954; Rancho Viejo, 1200 m. (2); total, 56 specimens.

Breeding choruses were noted throughout the lowlands in the summers of 1950 and 1951 after heavy rains filled the depressions and roadside ditches. In the dry season I have found *Smilisca* around springs, in the spathes of elephant-ear plants, and under water cans at Rancho Pano Ayuctle. None was found in bromeliads in the Cloud Forest. At Rancho del Cielo in 1949 the first *Smilisca* were heard on May 25, after a late afternoon thunderstorm. About 12 had gathered near a small water hole at the edge of the forest and were calling from saplings about 1 to 2 m. above the ground. Two nights later they were calling infrequently but synchronously, much like barnyard ducks quacking together. At Rancho del Cielo Frank Harrison found transforming frogs leaving his rain barrel around August 19, 1950.

Despite heavy rains none was heard at Rancho Viejo in late July, 1950, when I presume breeding activity had diminished.

At Gómez Farías one was recovered from the stomach of *Leptodeira septentrionalis*; at Rancho del Cielo, Heed found a *Leptophis* with one in its jaws.

Smilisca ranges north through the Tamaulipan Thorn Scrub to Texas; in addition it occupies Tropical Deciduous Forest and Cloud Forest in the Gómez Farías region.

Hyla eximia. — Known in Tamaulipas only from La Joya de Salas, 1550 m., (45). During the last two weeks of July, 1951, *H. eximia* appeared abundantly in the then lush meadows about the La Joya lake. Twenty-four were taken one afternoon in short wet grass, and 20 the following night while they were chorusing at one edge of the lake. Other small ponds in and near the open pine-oak woods of the La Joya valley were also inhabited. During the dry season when the valley is parched and overgrazed, and permanent water is scarce, we have found no trace of *H. eximia.*

In life these frogs were bright green above and white below. A dark stripe from the nostril through the eye to the femur is bordered above by a fine cream-white line.

H. eximia is a common inhabitant of wet meadows or savannas, usually associated with part of the oak-pine belt, that rim the dry central steppe of the Mexican Plateau. Tamaulipas is the northern outpost of the species on the east; however, I would expect records to appear from the oak-pine savannas of Nuevo León and Coahuila.

From *H. eximia* collected in Michoacán and Durango the La Joya series differs slightly in larger size (12 largest adults measure 29-32 mm. snout-coccyx) and a spotted, rather than striped or partly striped, dorsum.

Hyla staufferi. — Pano Ayuctle, 100 m. (4); Chamal, 150 m.; Gómez Farías, 360 m. (2); San Gerardo, 3 km. W, 120 m. (7); near the source of the Río Frío, 200 m. (6); total, 20 specimens. Smith and Taylor (1948) list one additional Tamaulipan record, Ciudad Mante, just outside the Gómez Farías region.

H. staufferi is abundant throughout the lowlands after heavy rains. Most of the specimens were taken at night when calling from low acacias and other thorny trees near temporary ponds. In the dry season I have encountered it twice, in the sheathing petioles of elephant-ear plants and around water cans, at Pano Ayuctle.

The distinctive call is low-pitched for a *Hyla*, a short "bought" or "ought." The color of living individuals is quite variable; during the day they are dull brown or olive drab above with stripes either gray green or dark brown, turning much paler at night. The vocal sac is orange yellow, the belly, cream white. *H. staufferi* ranges widely through the lowland tropics of Middle America, reaching its northern limit in southern Tamaulipas. In this area it occupies both tall Tropical Thorn Forest and Tropical Deciduous Forest at low elevations.

Hyla miotympanum. — La Union, *ca.* 120 m., Tropical Deciduous Forest, TU 15808 (5); Agua de los Indios, 4 km. SSW of Rancho del Cielo, Cloud Forest, 1300 m. (2); La Joya de Salas, *ca.* 3 km. SE, 1600 m., ?Dry Oak-Pine Forest, (26); total, 33 specimens.

In addition to the above, records from elsewhere in southern Tamaulipas include the Sierra de Tula, 2 km. SSE of Tula, 1300 m., dry pine woods, 110264; the Sierra de Tamaulipas, both near Acuña, 1000 m., dry oak-pine woods, and at Santa Bárbara, 400 m., heavy Thorn Forest.

It is remarkable that in a vertical range of 1500 m. through the Gómez Farías area only three collecting stations are represented, each close to permanent water or springs. Permanent sources of water were also a feature of the other Tamaulipan collecting localities. As noted above permanent springs in the karst (eastern) side of the Sierra Madre between the Río Sabinas and La Joya de Salas are very scarce, providing little habitat for amphibians breeding in such an environment. Thus the problem of dispersal between isolated water holes, separated from each other by 1 to 10 km. of very rough forested terrain, is puzzling. Either *H. miotympanum* is much less dependent on permanent water as a breeding site than present records indicate or else individuals indulge in considerable wandering during the wet season.

One specimen from the Sierra de Tula appeared about 3 m. above the ground in a bromeliad; I am unaware that any specimens from the Gómez Farías region were caught in bromeliads, although Walker obtained two at Indian Springs in low shrubs, about 1 m. above the ground.

Gastrophryne olivacea. — Three specimens from 4 km. west of San Gerardo, *ca.* 120 m., clearing in Tropical Deciduous Forest, comprise the only records from the Gómez Farías region. These were collected at night, two days after heavy rains on June 24, which brought a variety of lowland frogs to the breeding ponds. Just outside this area a recently transformed individual was collected under rocks at a pit near Xicotencatl, June 23, 1950. Reese and Firschein (1950) reported a singing male April 17, after a spring rain at Ciudad Mante. The southernmost record of which I am aware is one collected from a large chorus on July 13 near the Río Naranjos in a flooded pasture west of Antiguo Morelos, *ca.* 250 m.

There is no indication of an outer metatarsal tubercle in the present series, nor do these specimens show any suggestion of a middorsal band that might be expected if intergradation with *G. usta* occurred. Distributional overlap between these species in eastern Mexico should be sought; *usta* reaches middle Veracruz.

Compared with 16 *olivacea* from Texas and Oklahoma, Tamaulipan *Gastrophryne* differ slightly but consistently. They are more heavily mottled on the ventral surface of the groin, femur, and tibia. This mottling is intensified in the individual from the Naranjos Valley (110694) and extends to the abdomen, resembling many individuals of *G. carolinensis* in this feature. Whether the ventral mottling of Tamaulipan *Gastrophryne* represents an independent evolution or the result of Pleistocene contact with *G. carolinensis* remains undetermined.

Hypopachus cuneus. — Four records from clearings in Tropical Deciduous Forest of the Gómez Farías region include the following: Pano Ayuctle, *ca.* 100 m.; Chamal, 1 km. S, 200 m., TU 15495; Chamal, 2 km. E, 150 m. The last was plowed up by Dan Cameron in the summer of 1949. He reported often seeing small, pinkish frogs when plowing. The fourth individual was discovered in the stomach of *Coniophanes frangivirgatus*, collected near Pano Ayuctle in the late spring of 1950 (Darnell).

Rana pipiens. — Virtually ubiquitous wherever permanent streams or ponds occur, uncommon only in the karst areas, 78 specimens were taken from a variety of localities and habitats as follows: Río Boquilla near Chamal, (2); small stream at eastern foot of Mesa Josefeña, *ca.* 5 km. ESE of Pano Ayuctle, 200 m.; Río Sabinas between its source and Sabinas Bridge, 80 m. to 120 m., (15), TU 15481; water hole along trail from Pano Ayuctle to Rancho del Cielo, *ca.* 990 m. (9); Rancho del Cielo, 1050 m. (2), MMNH (20-6-54); permanent pond at Ojo de Agua de los Perros, 1 km. N of Agua Linda, 1900 m. (5); water hole at Carabanchel, 2000 m. (8); lake at La Joya de Salas, 1530 m. (31); San Antonio and vicinity, 850-900 m. (3).

During the rainy season *R. pipiens* probably wanders long distances into the mountains from such permanent lowland streams as the Río Sabinas and Río Guayalejo and thus succeeds in colonizing isolated pools, including a natural basin at Agua de los Perros and recently excavated cattle water holes near Rancho del Cielo and Carabanchel. In the bottom of a dry, vertical sinkhole about 9 m. deep near Rancho del Cielo I found a large adult, apparently trapped in its wanderings in the forest. One of very few suitable habitats not occupied is the allegedly permanent stream at Agua Linda which runs several hundred meters from a small cave spring before sinking underground. Agua de los Perros with its small *pipiens* population is only about a kilometer distant at slightly higher elevation in the same valley.

In May, 1953, at Carabanchel a small cattle tanque of roughly 20 m. diameter harbored a tremendous *R. pipiens* population. Harrell and I counted 85 frogs along 6 m. of shore line and estimated that the total number in the pond approached 1,000. The pond at this time was quite low, covered with a dense bloom of green algae; it probably dried out completely in the six virtually rainless weeks that followed.

The basin at Agua de los Perros contained a much smaller number, perhaps about 30, and Harrison's small artificial spring at Rancho del Cielo was never seen to hold more than ten adults despite the large number of tadpoles produced.

Ranging from the lowlands, at 80 m., almost to the top of the Sierra Madre at 2000 m., *R. pipiens* has a greater altitudinal distribution than any other species of reptile or amphibian in the Gómez Farías region.

Order Squamata
Suborder Sauria, Lizards

Hemidactylus turcicus. — Eight specimens were collected by R. Darnell in the town of Limón, 65 m.

Anolis sericeus. — Pano Ayuctle, 100 m. (41), TU 15512, 15449, 15445; La Union, 120 m. TU 15459, TU 15806 (3); Encino, 120 m., TU 15500; Gómez Farías, 360 m. (6); Chamal, 1 km. S, TU 15494; Mesa Josefeña, *ca.* 6 km. ESE of Pano Ayuctle, 300 m. (2); Valle de Paraíso, *ca.* 400 m.; 5 km. N of Ejido El Tigre (north of Ocampo), 930 m.; total, 60 specimens.

The record north of El Tigre is probably near the upper altitudinal limit of the species. A specimen from Acuña in the Sierra de Tamaulipas was obtained by H. Wagner at a similar elevation.

To my knowledge *A. sericeus* is mainly an inhabitant of lowland tropical forest, chiefly Tropical Deciduous Forest. None appeared in the Cloud

Forest or Oak-Pine Forests. The record above El Tigre came from an unusual area of dense, low, tropical evergreen shrubs with scattered trees, especially oaks *(Quercus germana)*. With few exceptions all specimens were captured on shrubs or in trees, both at night and during the day. Near Pano Ayuctle, Darnell found one at night on a lily pad *(Nymphaea)* in the Arroyo Encino about 3 m. from the nearest bank.

A. sericeus is a lowland tropical species throughout its range, reaching its northern limit in central Tamaulipas (Padilla, 13 km. SE, 90614). In life the males exhibited the blue gular spot in the center of an orange yellow fan characteristic of *A. sericeus*. The distributional gap separating *A. carolinensis* of eastern Texas from *sericeus* corresponds roughly to the arid corridor of B-type Köppen climates which lie in the Río Grande embayment. *A. carolinensis* is apparently derived from West Indian rather than Mexican *Anolis*.

Laemanctus serratus. — Pano Ayuctle and vicinity, 100 m. (14); La Union, 120 m., TU 15475, TU 15695; 2 km. NW of Pano Ayuctle, 330 m. (2); Gómez Farías, 350 m. (4); 3 km. WNW of Gómez Farías, 600 m. and 800 m. (3); total, 25 specimens.

Despite the size of this series few reliable habitat data are available. Most specimens were purchased by offering a bounty to inhabitants of Pano Ayuctle and Gómez Farías. I have seen only two alive in the field; one of these dashed across the Gómez Farías road in Tropical Deciduous Forest near the village, the other was found clinging to the trunk of a tree about 1 m. above the ground on the rocky limestone slopes above Gómez Farías (600 m.). The upper altitudinal limit probably lies above 800 m. as two specimens taken by Illoy Cordoba along the Rancho Viejo road supposedly came from near this elevation. I am not aware of any Cloud Forest records. Tropical Deciduous Forest of the lower mountain slopes and foothills is the characteristic habitat.

While climbing the steep trail from Pano Ayuctle to Rancho del Cielo on June 16, 1953, Walker and Harrell found two females in the act of excavating nests. A light but penetrating rain that morning had softened the hard-packed forest soil, and the lizards, found within a short distance of each other, were digging holes with their front feet. In an effort to escape both sought to reach trees, despite dense thickets of vines and brush along the trail that would have provided safety. In the collecting sacks one subsequently laid three eggs, the other five. Two of these measured 23 by 12 mm. and 22 by 12 mm.

In addition to Walker and Harrell's observation, evidence from gonad condition of preserved specimens also demonstrates that the beginning of the summer rains is correlated with reproductive activity. A large female collected January 1 contains large fat bodies with a minute ovary and unenlarged oviduct. Another collected March 4 is equally fat but with oocytes 3.5 mm. in diameter. With two exceptions, ten other adults (larger than 100 mm. snout-vent length) taken between late April and early July all have enlarged oocytes (larger than 10 mm. in diameter) or eggs enclosed in a tough leathery shell lying in the oviduct. A single juvenile, 41 mm. snout-vent length, was collected August 28.

Although occasional individuals are found on the ground, *Laemanctus serratus* is primarily arboreal. Two kept alive for a day at Pano Ayuctle jumped, hopped, or ambled along the ground, toadlike, without progressing fast enough to escape easy recapture. On a fence they seemed more at home and climbed nimbly, balancing with their tails. One was shot from a large fig tree inhabited by *Sceloporus serrifer* near Pano Ayuctle. The stomachs of six specimens yielded a snail shell and remains of an *Anolis*, in addition to a variety of arthropods, mainly beetles and Orthoptera.

Outside the Gómez Farías region I am aware of only one other Tamaulipan record, a single specimen taken near Morón, northwest of Tampico, by R. and J. Graber (University of Oklahoma collection). The distribution of *L. serratus* appears to correspond with that of dry tropical forests, Tropical Deciduous Forest, and, in the Yucatán, Thorn Forest. Possibly the species is not continuously distributed through extreme southern Veracruz and Tabasco where more humid Rainforest (inhabited exclusively by *L. longipes* and *L. deborrei*) is the predominant vegetation type.

Tamaulipan *L. serratus* differ from Yucatán and Campeche specimens in several features of scutellation. The name *alticornutus* is available for the latter; very probably *alticornutus* will prove to be a subspecies of *L. serratus*. I have not examined the types.

Ctenosaura acanthura. — Vicinity of Chamal, 150 m. (4); Río Frío 7 km. SE of Gómez Farías, *ca.* 140 m.; Ocampo, 5 km. NW, *ca.* 430 m.; Mesa Josefeña, 6 km. ESE of Pano Ayuctle, 360 m; near Pano Ayuctle, 100 m.; total, 8 specimens.

Arid and subhumid parts of the lowlands, especially arroyos in Tropical Deciduous Forest, characterize the habitat of the ctenosaur. The species is arboreal; several were shot in small oaks (near Chamal) and in other low trees. The individual from the Río Frío was shot in a cypress along the river. The Chamal and Ocampo valleys are local centers of abundance; in the Sabinas Valley near Pano Ayuctle *Ctenosaura* is decidedly scarce. An individual seen just west of the crest of the Sierra de Chamal (570 m.) gives the highest altitudinal record I have for the species. Another was observed above La Mula (560 m.) along the trail to Tula.

Ctenosaura is unknown north of central Tamaulipas; the northernmost stations are near Llera and Tepehuaje de Arriba, both close to the Tropic of Cancer.

Holbrookia texana. — Near the Río Frío, *ca.* 7 km. SE of Gómez Farías, 140 m. (19); Río Boquilla, SW of Chamal, 150 m. (2); Jaumave, 730 m., USNM 46725-6; total, 23 specimens. Loose, fine shale (Méndez) in dry areas of Thorn Forest characterizes the lowland habitats in which we found this species. It was quite numerous along the new irrigation canal cut near the source of the Río Frío. Here individuals sought concealment beneath scattered bushes and low brush.

In a recent revision Peters (1951) considered all Tamaulipan specimens to represent the subspecies *H. t. texana*. The Gómez Farías region is the southern limit of *H. texana* on the coastal plain. Records south of this latitude are from plateau localities.

Phrynosoma cornutum. — The only record within the Gómez Farías region is that of four collected by Wagner at 730 m. in Thorn Desert of the

Jaumave Valley. The dry lowlands between the Sierra de Tamaulipas and the Sierra Madre should be searched for additional records.

Sceloporus olivaceus. — Jaumave, 740 m. (5), USNM 46729, 46731-2 (Smith, 1939); *ca.* 12 km. S of Jaumave, 1000 m.; near Llera 60 km. N of Limón, *ca.* 270 m. (Smith, 1939); total, 10 specimens. There are also records for localities 3 km. E of Xicotencatl, *ca.* 200 m., and east of Llera at Hacienda La Clementina (Smith, 1939) just outside the Gómez Farías region. *S. olivaceus* is found in the Thorn Scrub and Thorn Desert of the coastal plain and the arid interior valleys. All that I have seen were in low trees. The absence of arid Thorn Scrub southward on the Gulf Coastal Plain probably limits this species in its lowland distribution.

Sceloporus grammicus. — Valle de la Gruta, 3 km. NW of Rancho del Cielo, 1500 m.; La Joya de Salas, 1510 m. (5); trail, Lagua Zarca and vicinity to La Joya de Salas, *ca.* 1350 to 2100 m. (17); Agua Linda and vicinity, 1800-1890 m. (7); trail above Montecristo *ca.* 10 km. NE of La Joya de Salas, 1700 m. (2); Carabanchel, *ca.* 2 km. S, 2040 m. (2); total, 34 specimens.

Ranging from 1350 to at least 2100 m., *S. grammicus* is one of the most abundant reptiles in the Humid Pine-Oak Forest of the Sierra Madre Oriental. In the Dry Oak-Pine Woodland near La Joya de Salas and at Carabanchel it is less numerous. The interesting pockets of Cloud Forest, surrounded by Pine-Oak Forest at Lagua Zarca, Valle de la Gruta, and elsewhere above 1400 m., are avoided by both this lizard and *S. torquatus.* Neither has been found in the main tracts of Cloud Forest below 1300 m.

The current trinomial arrangement of *S. grammicus* (Smith, 1939) in which all northern Mexican populations of this species are considered *g. disparilis* fails to account for variation apparent in the samples at hand. I have examined 106 specimens from six separate localities within the alleged range of *disparilis* and am able to separate each population on the basis of either dorsal scale number or head scale arrangement. In a series from the Sierra de Tamaulipas taken at 870 m., 20 specimens have a mean of 50.2 (range 47-53) dorsal scales. Mean of those from the Gómez Farías region 75 km. to the west is 65.3 (range 63-74, N = 27), and these individuals are smaller and darker.

Sceloporus serrifer. — Pano Ayuctle, 100 m. (27), TU 15497; *ca.* 5 km. ESE of Pano Ayuctle, 120 m.; El Encino, 110 m., TU 15498, 15501; 26 km. N of Limón, EHT 9411; total, 32 specimens. Tall trees, especially large strangling figs *(Ficus)* in lowland Tropical Deciduous Forest, including its edaphic modifications as palm forest and gallery forest, are occupied by this exclusively arboreal lizard. Three adults were collected and others were seen hiding among the stilt roots of a large fig tree left standing in a cornfield near Pano Ayuctle.

Lowland tropical forest of southern Tamaulipas is the northern range limit of *S. serrifer.* Not emphasized sufficiently in the type description (Martin, 1952) is the fact that Tamaulipan specimens, called *S. s. cariniceps,* resemble typical *serrifer* of Yucatán more closely than they do adjacent *S. s. plioporus* of Veracruz. That this similarity reflects a historical connection is doubtful; more likely, it is the result of parallel selective trends from wet to dry tropical forest in each area.

Sceloporus cyanogenys. — Rancho del Cielo, 1050 m. (26), MMNH (5-54); total, 27 specimens.

All of the above came from the eaves, roof, and walls of three wooden dwellings and adjacent stone fences in Mr. Harrison's Cloud Forest clearing. These constituted the driest and sunniest ground-level microhabitats available in an area surrounded, until recently, by tall, dark virgin forest. Suitable natural habitat in this largely forested region may be provided by the boulder piles on exposed ridges immediately north and south of the Rancho where I suspect *S. cyanogenys* occurs naturally. At Rancho del Cielo no observers have reported lizards of the genus *Sceloporus* in undisturbed Cloud Forest.

Typically *S. cyanogenys* is found in rocky terrain in the "brush country" of northeastern Mexico. Occasionally, it reaches higher altitudes, usually in pine-oak woods, as the Sierra de Tamaulipas (1000 m.), the Sierra Madre Oriental near Tula, 1400 m., and above Pablillo, Nuevo León, 2400 m. (Smith, 1939:223). Thus the Rancho del Cielo record is unusual but not altitudinally unprecedented.

In several minor morphological features Rancho del Cielo specimens differ from typical *S. cyanogenys.*

Sceloporus torquatus. — La Joya de Salas, 1510 m. (26); trail from Rancho del Cielo through Lagua Zarca to La Joya de Salas, specimens taken between 1350 m. and 2100 m. (23); Carabanchel and vicinity, 1950 m.; Agua Linda, 1800 m.; La Gloria, 2 km. WNW, 1650 m.; Valle de la Gruta 3 km. W of Rancho del Cielo, 1500 m.; trail between Montecristo and Carabanchel, 1630 m. to 1950 m. (4); total, 57 specimens.

Humid Pine-Oak Forest above 1300 m. is the center of abundance for this species. Drier Oak-Pine Forest and Woodland near La Joya de Salas and Carabanchel also provide suitable habitat. Along the La Joya de Salas-Rancho del Cielo trail below Lagua Zarca the local distribution of *S. torquatus* terminates abruptly at 1350 m. Here there is a sudden transition from sunny Pine-Oak Forest to deeply shaded Cloud Forest. Deficiency of sunlight at and near ground level may prevent *S. torquatus* from entering the Cloud Forest.

The pattern of distribution of the two common subspecies of this group, *S. t. torquatus* and *S. t. melanogaster*, is peculiar. Although they are allopatric, the first inhabits the eastern, southern, and apparently part of the western rim of the plateau, surrounding *melanogaster* on three sides. Despite a paucity of good ecological data on specific collections, this pattern may reflect a major ecological distinction between the two, i.e., completely collared *torquatus* selecting cool, relatively humid areas, especially pine-oak forest, and *melanogaster* the more arid central parts of the plateau (see Smith, 1936:576, for an interesting description of *melanogaster* in Zacatecas). Although intergrades between the two are reported in Michoacán (Smith, 1936), it is possible that strict ecological segregation isolates many other populations.

Recently four specimens of an unspotted, completely collared *S. torquatus* were taken in Pine-Oak Forest near Laguna del Progreso, 2400 m., *ca.* 40 km. NNW of El Salto, Durango (UMMZ 98998, 102576-8), the first record of a fully collared *torquatus* in the Sierra Madre Occidental north of Michoacán. Although Durango lizards have more femoral pores than has eastern *torquatus* (38, 39, 40, 41), their immediate derivation does not stem from *melanogaster*. They may represent a relic of an earlier Trans-Plateau emigration from the eastern Sierra Madre (see p. 90). In any case it illustrates the complex nature of the *melanogaster-torquatus* distribution.

North of Tamaulipas *S. t. binocularis*, an apparent *melanogaster* derivative, is reported in Nuevo León. Two specimens of *torquatus* (ssp. ?) from Miquihuana, Tamaulipas (Smith, 1936:574) have complete collars but otherwise resemble *S. t. binocularis* in low femoral-pore and dorsal-scale count, and thus may be intergrades between *S. t. torquatus* of the Gómez Farías region and *S. t. binocularis*.

The following data were taken on 20 specimens of both sexes from La Joya de Salas: femoral pores 27-37 (31.4); dorsal scales 27-31 (28.8); black nuchal collar three to five scales in width, complete in all specimens; maximum size, snout-vent 104 mm.

Sceloporus jarrovii. — La Joya de Salas and vicinity, 1500 to 1830 m.; 194 specimens. Since the majority of these were purchased from native collectors, the size of this series does not reflect an abundance of specific ecological data. All of those for which field notes are available were found beneath or on top of rocks, or were probed from cracks in boulders and in flat, bedded limestone. In the La Joya area a variety of plant communities within the Dry Pine-Oak Woodland and Chaparral zones are all occupied by *S. jarrovii*. Humid forest farther east in the Sierra Madre is avoided.

Except in the vicinity of La Joya, all four members of the *torquatus* group, *cyanogenys*, *serrifer*, *torquatus*, and *jarrovii*, are zonally isolated (see p. 83). In the La Joya valley the latter two were taken abundantly, but even here the zone of sympatry may not be wide. *S. torquatus* centers in more humid pine-oak areas east of the village, and *jarrovii* reaches its greatest abundance in drier habitats to the west. Field notes by Lidicker and Mackiewicz from July, 1951, are illuminating. They reported *S. jarrovii* most common in rocky areas north and west of La Joya, whereas *S. torquatus* was very common on rocks and trees in flat, open oak-pine savannas to the east. These grade into Pine-Oak Forest along the trail to Rancho del Cielo, an area not inhabited by *jarrovii*.

The present series resembles Hidalgan *S. j. immucronatus* in narrow width of the collar (two to three scales) and cobalt blue color in life. In 41 males the dorsal scale counts range from 38 to 47, mean 41.2 ± 0.7, and the femoral pores (both sides) from 28 to 42, mean 33.9 ± 0.9. In 57 females the dorsal scale range is 36 to 46, mean 40.5 ± 0.5, and the femoral pores vary from 26 to 39, mean 32.5 ± 0.7.

In life the dorsal color pattern of adult males is strikingly variable. Many individuals are a uniform steel gray or pale blue gray dorsally; others are bright blue, often with a pair of large, rusty red patches extending the full length of the body between fore and hind limbs and separated by three to four middorsal scale rows colored as the ground. Various degrees of patch development were noted, some specimens showing only a faint, tawny dorsal color on a steel-gray ground. In extreme development these patches resemble the color photograph of *S. j. erythrocyaneus* figured by Mertens (1950), differing only in their middorsal stripe.

North of the Gómez Farías region a specimen of *S. j. immucronatus* was extracted from a limestone ledge in oak chaparral east of Dulces Nombres, Nuevo León, 1440 m. Two others from Miquihuana, Tamaulipas, 50 km. SSW of the former, resemble *S. j. minor* with collars 5.5 and 6.5 scales wide (USNM 46741-2). Until adequate ecological and taxonomic studies on these and other members of the *jarrovii* complex are available, the distributional pattern will remain a puzzle. No intergrades have been reported to date, despite the fact that the entire range of *immucronatus* lies parallel to that of *minor*.

The distribution of *S. jarrovii immucronatus* fits the "Northeast Madrean" pattern nicely, although it is clearly less restricted to humid forest habitats than other members of this group.

Sceloporus variabilis. — Rancho Pano Ayuctle, 100 m. (128); La Union, 120 m., TU 15487, 15469; source of the Río Sabinas, 130 m.; Gómez Farías, 350 m. (30); Río Frío, *ca.* 7 km. SE of Gómez Farías, 100 m. (9); Sierra Madre Oriental W of Gómez Farías, 400-900 m. (3); Chamal and vicinity, 150-200 m. (6); slopes of Sierra Madre Oriental, *ca.* 2 km. WNW of Pano Ayuctle, 150-300 m. (3); Mesa Josefeña, 370 m. (3); San Antonio and vicinity, 820-910 m. (3); 10 km. WNW of Chamal, 450 m.; Ocampo, *ca.* 5 km. N, 360 m.; Aserradero del Refugio No. 1, 1050 m. (2); Rancho del Cielo, 1050 m. (5); La Joya de Salas, 1500-1550 m. (3); Jaumave, 730 m.; total, 201 specimens.

Below 1600 m. *S. variabilis* ranges throughout the Gómez Farías region, avoiding only humid, undisturbed forests. In areas such as the Cloud Forest and Tropical Semi-Evergreen Forest it frequents trails, clearings, road cuts, and other sunny places. Elsewhere in the lowlands *variabilis* is ubiquitous, ranging through the complete spectrum of xeric vegetation types of the Gómez Farías region, from the Thorn Scrub of the Jaumave Valley to the Tropical Deciduous Forest around Gómez Farías. No records above 1600 m. are known to me.

Although predominantly terrestrial, *S. variabilis* often climbs trees and walls of buildings.

I have not investigated population variation and possible subspecific affinity of collections in this large series.

Sceloporus parvus. — La Joya de Salas and vicinity, 1500-1750 m., 45 specimens. *S. parvus* occupies rocky areas with scanty underbrush in Dry Pine-Oak Woodland. Lidicker and Mackiewicz reported that most of their specimens were found in pairs on rocks in open pine woods north and west of La Joya.

Sceloporus scalaris. — La Joya de Salas, 1500 m. to 1750 m. Open rocky areas, occasionally with short grass, in Dry Pine-Oak Woodland of the La Joya valley were the source of these 28 specimens. No specific data are available to demonstrate ecological segregation between this and the other two small terrestrial fence lizards of this region, *S. parvus* and *S. variabilis.*

Hobart Smith (letter of March 10, 1955) kindly examined part of the present series and discovered no conspicuous difference between them and typical *S. s. scalaris* from farther south. They exhibit no trend toward *S. s. slevini*, which occurs near Pablillo, Nuevo León, 180 km. to the north. The present series was not taken in typical *slevini* habitat, bunch grass. More study of the *scalaris* group is needed; Smith's association of *slevini* with *scalaris* was admittedly tentative. There appear ecological as well as morphological grounds for associating *slevini* with *aeneus* rather than with *scalaris*.

Lepidophyma flavimaculatum. — Rancho del Cielo and vicinity, 1000-1100 m. (11), MMNH (26-5-54); trail between Rancho del Cielo and Agua de los Indios, *ca.* 1150 m.; below Lagua Zarca in Pine-Oak Forest, 1350-1500 m. (2); west slope of Sierra Madre, *ca.* 4 km. WNW of Lagua Zarca, 2150 m.; total, 16 specimens, all from Cloud Forest and Humid Pine-Oak Forest.

The statement that *Abronia* and *Lepidophyma* "are the only true cloud forest lizards" (Martin, 1955*a*) requires some qualification. Although I am unaware of records of *Abronia* from clearings, roadsides, or rock walls, *Lepidophyma* definitely inhabits such places. In heavy forest at some distance from clearings or natural openings this species is scarce. Along the narrow trail that formerly ran north of Rancho del Cielo through mature Cloud Forest to join the old "Company Road" one occasionally glimpsed a dark-brown lizard disappearing into cracks in large limestone boulders. None of these lizards was collected, but I assume that they were *Lepidophyma*. Unlike *Abronia*, which is more often, if not exclusively, found in forest *Lepidophyma* frequents both forested and cleared parts of the Cloud Forest zone. No other lizards are known in undisturbed forest of this region.

Most of the specimens listed above came from Rancho del Cielo, where they were captured under cover in wooden buildings and in a thick stone fence. Heed and Walker shot one in a fissure of a large boulder at night. In rather open forest near a small abandoned clearing west of Rancho del Cielo I discovered an adult quiescent beneath a large slab of loose bark on a dead tree, over 2 m. above the ground.

Of three specimens from Pine-Oak Forest above 1350 m., one was found under moist bark of a fallen pine, the second was lying quietly on leaf mold in a small depression, and the third was protruding its head from a small burrow. Two others seen in this same area escaped down well-marked burrows.

In a recent taxonomic account (Walker, 1955*a*) the present series is described as *Lepidophyma flavimaculatum tenebrarum*.

Leiolopisma silvicolum (?). — *ca.* 6 km. N of Gómez Farías at foot of the Sierra Madre, 120 m. (5), TU 15450; Mesa Josefeña about 6 km. ESE of Pano Ayuctle, 120-300 m. (6); Aserradero del Paraíso, 500 m; slopes north of El Tigre, *ca.* 1000-1050 m. (7); Rancho del Cielo, 1050 m., UMMZ 112916, MMNH (16-7-54); La Joya de Salas, 1500 m.; total, 23 specimens.

The majority of these specimens came from Tropical Deciduous Forest in the dry season when, despite the abundance of cover and evasiveness of the prey, as many as six were caught under rocks and dry leaves in a few hours. A road-constructing crew moving rocks and logs aided in assembling the series from north of El Tigre.

Leiolopisma occurs in Dry Oak-Pine Woodland as well as in tropical forest; specimens from the former habitat are indistinguishable morphologically from the lowland sample. In addition to those found at La Joya de Salas, one was obtained in dry pine-juniper woods in the Sierra de Tula at 1260 m. just southwest of the Gómez Farías region. Despite the Rancho del Cielo records I doubt that *Leiolopisma* was a Cloud Forest inhabitant until recent lumbering and other destructive activities modified this area. None was noted here or elsewhere in undisturbed humid montane forests prior to the summer of 1953.

A scarcity of specimens had hindered adequate taxonomic treatment of Mexican *Leiolopisma.* The status of the five members currently recognized in the *oligosoma* group, to which all Tamaulipan specimens are referred, is in need of revision. In the Smith-Taylor key (1950:157) males from the Gómez Farías region "key out" on the basis of leg length to *L. caudaequinae*, females to *L. gemmengeri forbesorum.* In a sample of 12 adults (UMMZ 101439-40, 111151, 111153-6, TU 15450) collected in tropical forests between 100 and 1050 m., sexual dimorphism in leg length is illustrated by the following. The adpressed limbs of five males measuring 45 to 50 mm. overlap by 4.8 scale rows (range 4 to 6); those of seven females between 49 and 52 mm. in snout-vent length are separated by a mean length of three scales (range 0 to 7).

Smith (1951:198) noted only one apparent instance of sympatry in the group. However, the data he gave for an individual considered *forbesorum* (EHT 23887) did not differ appreciably in any feature except leg length and possibly dorsal scale count from the data for a female paratype of *caudaequinae* (EHT 23886) taken at the same locality, west of Naranjo, San Luis Potosí. The apparent difference in leg length between the two is of the same order of magnitude as noted in the range of variation for adult females in the Gómez Farías region (see above).

On the basis of variation encountered in *Scincella (= Leiolopisma) silvicolum*, Darling and Smith (1954) reduced *caudaequinae* to subspecific rank under that name. Pending a thorough revision I use the combination *Leiolopisma silvicolum caudaequinae* with considerable reservation.

Variation in number of nuchal scales, from 1-0 to 3-3, and contact of the tertiary temporal with the parietal in many Tamaulipan specimens, point toward a close relationship with *L. laterale* and raise the question of intergradation between Mexican members of the *oligosoma* group and that species.

Eumeces tetragrammus. — Jaumave, 730 m.; Chamal, 2 km. E, 260 m.; Gómez Farías, 350 m.; total, 3 specimens. At Chamal, Dan Cameron referred to the individual he found in Palm Forest as a "glass snake" and seemed well acquainted with it. Other records from southern Tamaulipas include the Sierra de Tamaulipas near Acuña, 850 m.; Novilla Canyon, *ca.* 20 km. SW of Ciudad Victoria, 800 m.; and 5 km. E of Forlon (Smith and Taylor, 1950). These localities represent records in Thorn Desert and Tropical Deciduous Forest.

Eumeces dicei. — Aserradero del Paraíso, 500 m.; Rancho del Cielo,

1060 m. (10); *ca.* 1 km. N of Rancho Viejo, 1320 m.; lower part of Pine-Oak Forest E of Lagua Zarca, 1450-1500 m. (2); 1 km. W of Lagua Zarca, 1850 m.; Agua Linda, 1800 m. (5); La Joya de Salas, along trail to about 4 km. SE of the village, 1550-1600 m. (7); total, 27 specimens.

E. dicei appears to be endemic to southern Nuevo León and Tamaulipas. It has an interesting distribution. In addition to the type locality, other southern Tamaulipas records include the Sierra de Tamaulipas near Acuña, Oak-Pine Woodland, 900 m. (3), and Chihue, NW of Ciudad Victoria, Pine-Oak Forest, 1860 m. Although the species ranges down to 540 m. (type locality) and 500 m. (Aserradero del Paraíso), the majority of specimens from the Gómez Farías region were obtained in Pine-Oak Forest. It should be sought in similar habitat southward in San Luis Potosí for possible fit to the "northeast Madrean" pattern.

The series of 10 from Rancho del Cielo all came from a small abandoned clearing about 1 km. west of the Rancho. Like *Leiolopisma*, *Sceloporus variabilis*, and *S. cyanogenys* this lizard is probably not part of the undisturbed Cloud Forest fauna. Present lumbering should promote its spread through much of this region.

E. dicei is a short-legged skink and, to my knowledge, exclusively terrestrial.

Ameiva undulata. — Pano Ayuctle, 100 m. (30), TU 15496, 15510, 15425; La Union, 120 m., TU 15805, 15462 (6); Gómez Farías, 350 m. (2); Río Boquilla, SW of Chamal, 150 m.; total, 43 specimens.

This wide-ranging species of the tropical lowlands apparently avoids Thorn Forest, the latter occupied by *Cnemidophorus*, and selects instead Tropical Deciduous Forest. The only locality where I am certain this species occurs outside Tropical Deciduous Forest, and the highest elevation recorded for the species in Tamaulipas, is oak woodland with scattered palmettos in the Sierra de Tamaulipas near Acuña, 900 m.

Smith and Laufe (1946) listed only two records from north of the Gómez Farías region, Hacienda la Clementina near Llera and Ciudad Victoria, to which I can add a third, 15 km. NE of Zamorina, *ca.* 150 m. None of these records greatly exceeds the known northern extent of Tropical Deciduous Forest in southern Tamaulipas.

Tamaulipan specimens are referable to the subspecies *A. u. podarga.*

Cnemidophorus sacki. — Pano Ayuctle, 100 m. (14); Encino, 120 m., TU 15471, 15519 (9); Río Frío, *ca.* 7 km. SW of Gómez Farías, 120 m. (7); Gómez Farías and vicinity, 350 m. (2); source of the Río Sabinas, 140 m.; Chamal, 150 m. (2); Ocampo, 5 km. NW, 330 m. (2); San Antonio and vicinity, 910 to 1100 m. (3); Jaumave, 800-1000 m. (27); total, 68 specimens.

As is obvious from number of specimens and the localities represented above, *C. sacki* is quite common in arid and subhumid parts of the Gómez Farías region, including Thorn Desert, Thorn Forest, and Tropical Deciduous Forest. The maximal elevation attained here is 1100 m. east of San Antonio in a savanna of widely scattered oaks and palmetto clumps. None was noted in Dry Oak-Pine Woodland near La Joya, although *sacki* ranges into this vegetation type in the Sierra de Tamaulipas (1000 m.).

The problem of ecological distinction between *Ameiva* and

Cnemidophorus needs careful study. *Ameiva* avoids most of the drier lowland habitats such as Thorn Scrub; the two occur together mainly in areas of Tropical Deciduous Forest. Only *Cnemidophorus* has been found at moderate elevations in the interior valleys.

All of the specimens are referable to *C. s. gularis.*

Abronia taeniata. — Rancho del Cielo and vicinity, 1000 to 1500 m. (22), MMNH (1-5-54, 20-6-54); Valle de la Gruta, *ca.* 3 km. WNW of Rancho del Cielo, 1500 m.; Aserradero de Socorro, *ca.* 5 km. SW of Rancho del Cielo, 1470 m.; Rancho Viejo, 1200 m.; Lagua Zarca and vicinity, 1350-2000 m. (4); total, 31 specimens. A large part of the series designated Rancho del Cielo and vicinity was purchased from sawmill workers in the spring of 1953, at a time when road construction and tree cutting were under way in Cloud Forest and Humid Pine-Oak Forest 1 to 3 km. west of Rancho del Cielo between 1000 and 1500 m. The presumption that specimens came from this area does not exclude the possibility of a few originating elsewhere in the Gómez Farías region. All 17 specimens for which field data are available were found in Cloud Forest and Humid Pine-Oak Forest from 1000 to 2000 m. on the east side of the Sierra Madre Oriental. This habitat preference and the known range of the species fit the northeast Madrean pattern (see p. 88). There are no records between Xilitla, San Luis Potosí (Taylor, 1953) and Tamaulipas. For further discussion of habits and systematic status, see Martin (1955*a*).

Gerrhonotus liocephalus. — Near Gómez Farías, 300 m.; 1 km. WNW of Pano Ayuctle, at foot of the Sierra Madre, *ca.* 150 m.; *ca.* 5 km. N of El Tigre 930 m.; Lagua Zarca and vicinity, 1400-1600 m. (3); total, 6 specimens.

As indicated by the above localities *Gerrhonotus* is more diverse in ecological range than most members of the Gómez Farías region fauna and is one of few species found both in lowland Tropical Deciduous Forest, Tropical Evergreen Forest, and Humid Pine-Oak Forest. This wide ecological amplitude is paralleled by certain differences in morphology.

Gerrhonotus and *Abronia* were collected along the same part of trail between the upper edge of the Cloud Forest and the Lagua Zarca doline; elsewhere they are not known to occur together.

Outside the Gómez Farías region Walker and Harrell collected two juveniles under a log at 1500 m., in mixed pine-oak-sweet gum forest along the Dulces Nombres road NW of Ciudad Victoria. The individual reported by Tihen (1948) from near Ciudad Maiz came from immediately south of the Gómez Farías region.

One of the diagnostic features of *G. l. loweryi,* the second primary temporal in contact with the fifth medial supraocular, is evident in all eight Tamaulipan specimens. Other features, however, including low number of caudal whorls (135 and 140 in UMMZ 98983 and 101296), and low number of dorsal scale rows, indicate a trend toward *G. l. infernalis.* The number of dorsal scale rows may vary with altitude and habitat. The following counts were obtained: Pine-Oak Forest near Lagua Zarca, 1400-1600 m., 50, 51, 53; similar forest northwest of Ciudad Victoria, 1500 m., 49 and 52; lowland tropical forest below 900 m., 58, 58, 62. Further morphological studies on this species should therefore include altitudinal subdivision where sample size and field data permit. Some of the variation within the type series of *loweryi* from the Xilitla region (taken at different elevations?) may conceivably be attributed to this factor.

Xenosaurus newmanorum. — La Joya de Salas, 4 km. SE, 1700 m; La Union, 150 m. (?), TU 15473 (2). These two localities represent Dry Oak-Pine and Tropical Deciduous Forest localities, respectively, a remarkably diverse ecological range, but one that is shared by several other species, such as *Leiolopisma.* For further discussion of systematic and distributional status see Martin (1955*a*).

Suborder Serpentes, Snakes

Leptotyphlops myopicus. — Pano Ayuctle, 100 m. (2); Gómez Farías and vicinity, 360 m.; Aserradero del Paraíso, 420 m.; Jaumave, 735 m.; San Antonio, 900 m.; total, 6 specimens.

Although only six specimens were taken in the Gómez Farías region, they represent a wide range of habitats from Tropical Evergreen Forest (Aserradero del Paraíso) to Thorn Desert (Jaumave). Of the two specimens from Pano Ayuctle, one was found dead in the Río Sabinas, the other appeared from beneath a camp stove just after dark. The San Antonio specimen was taken on the ground along a trail at night.

Leptotyphlops reaches oak-pine woodland at 900 m. in the Sierra de Tamaulipas.

Constrictor constrictor. — Pano Ayuctle, 100 m. (2); Río Frío, SW Gómez Farías, 110 m. (2). This boa, often called "navaja," is generally well known to residents of the tropical lowlands. In southern Tamaulipas it has not been found north of Tropical Deciduous Forest, its typical habitat in the Gómez Farías region.

Adelphicos quadrivirgatus. — Low Tropical Evergreen Forest, 13 km. N of Ocampo, 900-1000 m. (11). At this locality *Adelphicos* and *Tantilla rubra* were fairly common under stones and other cover. With the help of a 30-man road-building crew the above series was obtained in a day and a half. A brief taxonomic discussion of these specimens recently appeared (Martin, 1955*a*); they are referred to *A. quadrivirgatus newmanorum.*

Amastridium sapperi. — Cloud Forest at Rancho del Cielo, 1050 m. (2). A juvenile of this rare snake appeared on the flagstone floor of a recently constructed building at the Rancho. The other, an adult female, was collected by Frank Harrison on carpet grass in the Rancho del Cielo clearing.

The striking geographic distribution of *A. sapperi,* unknown between northeastern Mexico and the Pacific foothills Rainforest of Chiapas, is not matched by any other terrestrial vertebrate in the Gómez Farías region fauna. For further distributional and taxonomic discussion see Martin (1955*a*).

Coluber constrictor. — *ca.* 2 km. E of San Antonio, 1060 m.; low palmetto grassland, a transitional, treeless belt between Thorn Scrub and Dry Pine-Oak Woodland.

South of the Río Grande *Coluber* is recorded from only five localities including one in the Petén of Guatemala and four in Mexico (Ethridge, 1952). The present individual divides a distributional gap of 600 km.

between Matamoras, Tamaulipas, and Tecolutla, Veracruz. At present it would appear that the rarity of this genus in Middle America reflects a relic pattern, presumably the outcome of Pleistocene contact between temperate and tropical savannas. Certainly *Coluber* is much less abundant in Mexico than its close relatives, *Masticophis* and *Dryadophis*, which are continuously distributed. Admittedly, the exact range of *Coluber* and the ecological basis for its present distribution remain undefined.

The following scale counts were obtained on the San Antonio specimen: supralabials 7-8, infralabials 7-7, scale reduction formula $17 \frac{-4(77)}{-4(87)} 15$ (158), 97 caudals. The individual is an immature male, total length 484 mm., body length 345 mm. There is no suggestion of the juvenile color pattern. Evidently it represents *C. constrictor stejnegerianus*.

Coniophanes frangivirgatus.— Pano Ayuctle, 100 m., Tropical Deciduous Forest. No additional specimens have appeared despite continued collecting at this locality since spring of 1950 when the above individual was taken (Martin, 1955*a*).

Coniophanes imperialis. — Pano Ayuctle, 100 m. (5); Encino, 120 m.; *ca.* 3 km. NW Gómez Farías, TU 15524; 4 km. NW of Chamal, elevation *ca.* 150 m., TU 15522. Of a total of eight specimens three were caught in sugar-cane fields. Areas presently or formerly covered by Tropical Deciduous Forest characterize the above localities. In the Sierra de Tamaulipas *C. imperialis* attains an elevation of 900 m., 3 km. S of Acuña in open oak woods (101221).

The present series exhibits the middorsal stripe of *C. i. imperialis*.

Dryadophis melanolomus. — Pano Ayuctle, 100 m. (2); Gómez Farías, 350 m.; *ca.* 3 km. WNW of Pano Ayuctle, 300 m.; Aserradero del Paraíso, 450 m.; Rancho del Cielo, 1050 m.; total, 6 specimens.

The more mesic of the lowland and foothill tropical habitats including tall Tropical Deciduous Forest and Tropical Evergreen Forest were the sources of all but one of these six specimens. The latter allegedly came from a sawmill near Rancho del Cielo, but this locality information is not reliable. I am unaware of coastal plain records at any distance from the Sierra Madre; the northern range of *Dryadophis* corresponds with the limit of tall Tropical Deciduous Forest and Evergreen Forest of the lower Sierran slopes.

Northeast Mexican specimens are considered *D. m. veraecrucis* (Smith and Taylor, 1945).

Drymarchon corais. — Pano Ayuctle, 100 m. (9), TU 15812; vicinity of Gómez Farías, 350 m.; La Union, 120 m., TU 15682; Rancho del Cielo, 1050 m.; total, 13 specimens.

The record from Rancho del Cielo allegedly came from the hillside immediately north of Frank Harrison's clearing. It was collected in late 1953, sometime after lumbering of the area, and probably represents a recent invader of the Cloud Forest. Tropical Deciduous Forest and, in the Sierra de Tamaulipas, Oak Savanna, characterize other localities in southern Tamaulipas where *Drymarchon* has been taken.

Two of the Pano Ayuctle specimens were found swimming across the Río Sabinas.

Drymobius margaritiferus. — Pano Ayuctle, 100 m. (7); Chamal, 4 km. NW, *ca.* 150 m., TU 15527; Aserradero del Paraíso, *ca.* 420 m. (3); Gómez Farías and vicinity, 300-350 m. (3); road between Gómez Farías and Rancho del Cielo, *ca.* 750 m.; Rancho del Cielo and vicinity, 1050 m. (6); San Antonio, 900 m.; total, 22 specimens.

In addition to four specimens designated Rancho del Cielo and obtained in 1953 from sawmill workers, one was collected on dry leaves in heavy forest in 1948, establishing the natural occurrence of *Drymobius* in the lower part of the Cloud Forest. The rest of the 22 specimens were found in a variety of tropical habitats, both arid and humid. *Drymobius* reaches 900 m. in pine-oak woods in the Sierra de Tamaulipas.

Elaphe triaspis. — Gómez Farías, 350 m.; between Gómez Farías and Rancho del Cielo, elevation unknown; between Rancho del Cielo and Pano Ayuctle, *ca.* 500 m. (2); total, 4 specimens.

None of the three species of *Elaphe* from the Gómez Farías region is well represented in collections. The four cited above are from montane tropical slopes (Tropical Semi-Evergreen Forest) below Cloud Forest. One record south of this area is from the Sierra Cucharas at *ca.* 300 m., 8 km. NNE of Antiguo Morelos.

The name *E. triaspis intermedia* is used for Mexican populations north of Tehuantepec formerly called *E. chlorosoma* (Dowling, 1952 *b*).

Elaphe flavirufa. — Pano Ayuctle, 100 m. (4); La Union, 120 m., TU 15483; Aserradero del Paraíso, *ca.* 420 m.; total, 6 specimens.

Dowling (1952*a*) has pointed out that despite certain convergent trends, *E. flavirufa* and *E. guttata* are not known to intergrade in the narrow zone of overlap between the two species in southern Tamaulipas and eastern San Luis Potosí. In the region of sympatry I suspect *E. flavirufa* is more often found in humid forests (Tropical Semi-Evergreen Forest) than *E. guttata*. Tamaulipan material studied by Dowling was considered *E. f. flavirufa*.

Elaphe guttata. — La Union, 120 m., TU 15484; Chamal, 2 km. E, 150 m.; Sabinas Bridge, *ca.* 20 km. N of Limón, 90 m.; Jaumave, 740 m. (2); total, 5 specimens from Thorn Desert and Tropical Deciduous Forest.

Two other records immediately south of the Gómez Farías region, 5 km. south of Ciudad Mante and *ca.* 8 km. west of Antiguo Morelos, 480 m., represent regions of Tropical Deciduous Forest. *E. guttata* follows arid tropical forests (?Tropical Deciduous Forest) at least as far south as Ebano, San Luis Potosí (Taylor, 1952).

Dowling (1951) treated populations formerly considered *Elaphe laeta laeta* as *E. guttata emoryi*.

Ficimia olivacea. — Pano Ayuctle, 100 m. (3); La Union, 120 m., TU 15465, 15477; Rancho del Cielo (?), 1050 m.; total, 6 specimens. When I inquired about it early in 1953, Frank Harrison seemed fairly certain that the latter was collected at Rancho del Cielo in November of 1952. Until an authentic record appears, I do not consider this sufficient to include *Ficimia* as part of the Cloud Forest fauna. The other localities represent Tropical Deciduous Forest.

One of the Pano Ayuctle specimens was found dead on a trail at night, another was dug up an inch below the surface of the ground in an abandoned clearing, and a third appeared in the stomach of a coralsnake, *Micrurus*.

This series is quite variable with regard to dorsal blotches, and two individuals are unicolor. I see no basis for following Taylor (1949) in retention of *streckeri* as a species. In addition to color pattern his data on a large series from Xilitla clearly show intergradation between *streckeri* and *olivacea* in postocular count. Tamaulipan specimens are referred to *F. o. streckeri.*

Geophis semiannulatus. — Rancho del Cielo and vicinity, 1050 m. (7); Rancho Viejo, 1200 m. (2); Agua Linda, 1800 m. (3); total, 12 specimens.

In its apparent restriction to humid montane forest (both Cloud Forest and Pine-Oak Forest in the Gómez Farías region) and its occurrence at Guerrero, Hidalgo, the only other authentic locality, *G. semiannulatus* matches the "Northeast Madrean" pattern. Variation and local distribution of the present series have been discussed elsewhere (Martin, 1955*a*).

Imantodes cenchoa. — Pano Ayuctle, 100 m. (2); source of Río Frío, *ca.* 5 km. SSE of Gómez Farías, *ca.* 150 m.; all from the Tropical Deciduous Forest zone.

These three constitute the only records of *Imantodes* from Tamaulipas. Two were found in the eaves of a palm thatch hut. All are identified as *I. c. leucomelas* (Martin, 1955*a*).

Leptodeira septentrionalis. — Pano Ayuctle, 100 m. (2); La Union, 120 m., TU 15464, TU 15485, TU 15513; Gómez Farías, 350 m. (2); Aserradero del Paraíso, 420 m.; 3 km. WNW of El Carrizo, east flank of the Sierra Gorda, 450 m.; Rancho del Cielo, 1050 m. (7) MMNH (1); La Joya de Salas, 1500 m. (2); total, 19 specimens.

The altitudinal and ecological range of *L. septentrionalis* exceeds that of any other member of the ophidian fauna in the Gómez Farías region. The only major vegetational types from which specimens are unknown are Humid Pine-Oak Forest and lowland and interior basin Thorn Desert and Thorn Scrub. Most of the specimens labeled Rancho del Cielo were caught in buildings or on fences, rather than in the forest itself. *Leptodeira* may not be a true member of the Cloud Forest fauna.

This series represents *L. s. septentrionalis.*

Leptodeira maculata. — Pano Ayuctle 100 m. (8); La Union, 120 m., TU 15461; San Gerardo, *ca.* 5 km. W; Chamal, within a radius of 4 km., 150 m. (3), TU 15820, TU 15529; *ca.* 8 km. E of Chamal, 120 m. (2); total, 17 specimens from Tropical Deciduous Forest.

In contrast to *annulata, L. maculata* is restricted to the tropical lowlands, invariably near streams and irrigation ditches. The two species were collected together at Pano Ayuctle, La Union, and outside the Gómez Farías region at Santa Bárbara, 510 m., in foothills of the Sierra de Tamaulipas. At the latter site I captured one *annulata* and four *maculata* on the same night at a water hole and small surface stream. This is the highest altitudinal record of *maculata* known in southern Tamaulipas.

The details of possible competition in those lowland areas where both occur together remain unknown.

I am indebted to William Duellman for checking all identifications of *Leptodeira.*

Leptophis mexicanus. — Pano Ayuctle, 100 m. (6); Limón, 5 km. N, 80 m.; Chamal and vicinity, 150 m. (4); Gómez Farías, 350 m. (2);

Rancho del Cielo and vicinity, 900-1100 m. (7), MMNH (26-5-54); total, 21 specimens.

Tropical Deciduous Forest and lower sections of the Cloud Forest include the local range of *Leptophis*. The disappearance of these habitats northwards probably controls the distribution of this species at its northern limit. Although three of the Rancho del Cielo records lack specific locality data and three others came from clearings or partially lumbered areas, one, UMMZ 101365, was definitely taken in undisturbed Cloud Forest.

Masticophis flagellum. — Pano Ayuctle, 100 m.; La Union, 120 m. TU 15482; Sabinas Bridge, 18 km. N of Limón, 100 m.; Chamal, 2 km. E, 150 m.; Jaumave, 730 m.; total, 5 specimens.

Other specimens immediately south and east of the Gómez Farías region were found dead on the highway, 5 to 11 km. south of Ciudad Mante, and 3 km. northeast of Xicotencatl. One captured near Chamal was in an open brushy area of acacias at the edge of a dense palm forest.

M. flagellum and *M. taeniatus* appear completely sympatric in their lowland distribution. Both species inhabit Thorn Forest and low Tropical Deciduous Forest. Only *M. taeniatus* was found at higher elevations (La Joya de Salas).

Specimens from the Gómez Farías region resemble *M. f. testaceus*.

M. taeniatus. — 2 km. S of El Carrizo, *ca.* 300 m.; 25 km. N of Limón, 120 m.; Jaumave, 800-1000 m.; La Joya de Salas, 1500 m.; 4 specimens.

The lowland and Jaumave specimens are considered *M. taeniatus ruthveni*; the individual from La Joya is definitely not *ruthveni* and exhibits several of the characters of *schotti*. The lowland distribution of both *taeniatus* and *flagellum* does not exceed those areas of Thorn Savanna extending down drier parts of the Gulf Coastal Plain into northern Veracruz. In the interior at higher elevation both follow arid environments farther south.

Oxybelis aeneus. — Pano Ayuctle 100 m.; Gómez Farías, 350 m.; trail between Gómez Farías and Rancho del Cielo, *ca.* 480 m. Three specimens from areas of tall Tropical Deciduous Forest and Semi-Evergreen Forest constitute the only records of this tree snake.

Pliocercus elapoides. — Cloud Forest at Rancho del Cielo and vicinity, all within a radius of about 3 km., 1000-1250 m. (46); Rancho Viejo, 1200 m. (2); total, 48 specimens.

Most reptiles in the humid forests of the Gómez Farías region are more common about clearings and trails than they are in the forest itself, and many species were found only in such places. *Pliocercus* is one of few which do not concentrate in clearings; of 15 specimens accompanied by specific habitat data only two were taken in the Rancho del Cielo clearing.

Although the species reaches 2000 m. on Mount Ovando in Chiapas (Smith, 1941) and 1000 m. in Alta Verapaz (Stuart, 1948), it is largely confined to humid lowlands and foothills on the Atlantic slope of Mexico. In Tamaulipas, however, only the type of *P. e. celatus* is reported from below 1000 m. (Ciudad Victoria, elevation 320 m., coll. by Weldon Embury). I have sought without success to obtain further information concerning this record. Possibly the specimen came from humid montane forest immediately west of Victoria; the general range of the genus in humid tropical environments, and its apparent salamander-feeding habit (Stuart, 1948:72) make a record from the arid plains about Victoria appear improbable. On the other hand, *P. e. celatus* definitely occurs at low elevations further south, i.e., two miles south of Tihuatlan, Veracruz (Smith, Smith, and Werler, 1952).

Indirect evidence points to the summer rainy season as the time of egg-laying. A female captured June 15, 1948, laid three of a complement of at least five eggs soon after capture. Of a clutch of eight eggs found in a compost heap by Frank Harrison about July 22, 1950, five had either hatched or been destroyed and the remaining three were in an advanced state of development when examined on August 14. A second set of six was found in a rotten stump in heavy forest on August 16. It included two empty shells and one embryo about to hatch, with black and red bands fully developed and the yellow bands indicated but still very pale.

Of many stomachs examined only one contained remains of food, a salamander of the genus *Pseudoeurycea.*

In the Gómez Farías region *Pliocercus* is one of three reptiles and amphibians that are unknown outside Cloud Forest. The absence of a well-developed vertebrate fauna restricted to Cloud Forest (Subtropical Zone) contrasts with the unique vegetational and floristic assemblage found between 1000 m. and 1450 m. in the Rancho del Cielo region (Martin, 1955*b*). More information on the local range of *Pliocercus* is needed. If it feeds largely on salamanders it should appear in the Humid Pine-Oak Forest.

The present series is considerably larger than any other from a single locality. It should clarify the nature of population variation and the status of *P. e. celatus.*

Rhadinaea crassa. — Rancho del Cielo and vicinity, (27), MMNH (19-5-54); sawmills of the Rancho del Cielo area, exact locality unknown, probably between 1000 and 1300 m. (32); trail to Agua de los Indios, *ca.* 3 km. SW of Rancho del Cielo, 1200-1300 m. (2); Valley de la Gruta, 3 km. W of Rancho del Cielo, 1500 m.; Casa Piedras and vicinity, *ca.* 1 km. N of Rancho Viejo, 1290-1350 m. (4); Aserradero Refugio No. 2 de Oton Diaz, 1680 m. (2); North Woods, *ca.* 4 km. N of Rancho del Cielo, 1380 m. (2); trail, Rancho del Cielo to La Joya de Salas, 1100-1750 m. (2); Lagua Zarca, 1590 m.; Agua Linda, 1800 m. (4); La Joya de Salas, 4 km. SE; total, 78 specimens.

Represented by specimens from a variety of localities, *Rhadinaea* is virtually the only snake in the Gómez Farías region that is common enough to restrict zonally with some confidence. It is found through the belt of humid montane forest between 1000 and 1800 m., mainly on the eastern side of the Sierra Madre. Individuals were found in all types of terrestrial habitats, on trails, truck roads, dry leaves of the forest floor, grass in clearings, inside or under logs, beneath stones or bark, and in one instance about 10 m. inside the mouth of a cave (Walker and Heed). Cloud Forest and Humid Pine-Oak Forest are the typical habitats; the La Joya record indicates that drier oak-pine woods may also be occupied.

Salamanders may constitute the major item of diet. Although the stomachs of most specimens were empty, Walker forced one to regurgitate a *Pseudoeurycea scandens*, and another ate the tails of two *P. scandens* which were confined with it. Parts of three *Chiropterotriton* were found in stomachs of two other specimens, with a *Pseudoeurycea* tail and complement of 12 salamander eggs enclosed in a jelly mass in the stomach of a third. This is the only direct evidence of salamander breeding that has come to light in this area to date. Unfortunately, the exact date of capture of the snake is unknown although it probably was midsummer of 1953.

South of the Gómez Farías region *Rhadinaea crassa* is reported from a few localities at moderate elevation in Hidalgo and San Luis Potosí. It is a member of the "Northeast Madrean" faunal group.

R. crassa is probably conspecific with *R. gaigeae*. A paratype of the latter (UMMZ 90668) was collected in 1879 and is greatly faded; however, it clearly shows black spots on the ventral tips, one of three features Smith (1942) considered diagnostic of *crassa*. *R. montana* of Nuevo León also is quite possibly conspecific.

Rhinocheilus lecontei. — South of Ciudad Victoria, 24 km. N of Llera, AMNH 72404. On the coastal plain the long-nosed snake cannot be expected very far south of this point. Like many other members of the "Arid Interior" component, *Rhinocheilus* ranges slightly farther south on the Plateau than on the coastal plain, reaching central San Luis Potosí (Taylor, 1953). Another desert and arid interior genus, *Hypsiglena*, occurs near the Gómez Farías region at Soto la Marina (101220) and Hacienda la Clementina.

Salvadora lineata. — La Joya de Salas, *ca.* 1600 m.; San Antonio, 2 km. E, 1060 m. The latter was found in a grassy savanna of scattered oaks and palmettos. Lowland records should be sought in the drier parts of the coastal plain and mesas. The southernmost specimen recorded from the coastal plain in the Museum of Zoology, University of Michigan, is from Soto la Marina. On the plateau this species follows arid habitats farther to the south.

Scaphiodontophis cyclurus. — Gómez Farías, 350 m. This is the only record of the genus north of central Veracruz (Martin, 1955*a*); the specific identification is tentative.

Spilotes pullatus. — Pano Ayuctle, 100 m. (6); Gómez Farías, 350 m. (3); Chamal, 200 m.; total, 10 specimens from Tropical Deciduous and Semi-Evergreen Forest. In addition to the above specimens *Spilotes* was seen on several occasions along the lower slopes of the Sierra Madre in heavy forest up to 450 m. Harrell and Walker noted one pursuing a mouse at this elevation on the trail between Pano Ayuctle and Rancho del Cielo. Others near Pano Ayuctle were taken in cultivated fields or in brush near the Río Sabinas; one was found swimming in the river. A female collected April 11, 1953, contained 11 large oviducal eggs.

Spilotes is a distinctive member of the lowland tropical fauna in eastern Mexico, ranging from southern Tamaulipas to Argentina. It is unknown north of Tropical Deciduous Forest in the Gómez Farías region.

No taxonomic review was made of this material; currently all Mexican *Spilotes* are assigned to *S. p. mexicanus*.

Tantilla rubra. — Vicinity of Gómez Farías, *ca.* 350 m.; Rancho del

Cielo and vicinity (?), (4); N of El Tigre, 1000-1050 m. (8); total, 13 specimens.

Three of the Rancho del Cielo records were sawmill specimens and the fourth was obtained by Illoy Cordoba of Gómez Farías, who reported finding it in the Cloud Forest. As the exact source of these specimens is unknown, the occurrence of *Tantilla* in the Cloud Forest is not definitely established. The individual from Gómez Farías had been swallowed by a coral snake. The series taken north of Ocampo all were found below flat, bedded limestone rocks in an area of dense, low tropical thickets with scattered trees.

In addition to Tropical Semi-Evergreen Forest and possibly Cloud Forest which *Tantilla* inhabits in the Gómez Farías region, one was captured at Chihue, northwest of Victoria, *ca.* 1860 m., an area of Humid Pine-Oak Forest. The northern limit of *Tantilla rubra* extends at least as far along the Sierra Madre as Horsetail Falls, Nuevo León (Smith, 1944).

Near Tehuacan, Puebla, *T. rubra* is found in much drier habitat (Smith, 1943).

Trimorphodon tau. — Dry Pine-Oak Woodland at La Joya de Salas, 1550 m.; Thorn Forest 20 km. N of Llera, AMNH 72399-72400.

Smith and Darling (1952) recently added this genus to the Tamaulipan fauna. On the basis of variability of head pattern, they consider *upsilon* conspecific with *tau.* In addition to the individual from La Joya de Salas I have seen another specimen from south of Ciudad Victoria and found an *upsilon* type pattern in the latter and a *tau* type in the former. It seems improbable that this slight distinction reflects the existence of two separate species, and I follow Smith and Darling in considering them to be the same.

Trimorphodon tau is widespread through the Central Plateau. Outside Tamaulipas it is unknown from the Gulf Coastal Plain.

Tropidodipsas fasciata. — Vicinity of Gómez Farías, *ca.* 350 m., UMMZ 110988. Eight snakes preserved by Illoy Cordoba of Gómez Farías between March 13 and April 5, 1953, included *Leptotyphlops, Leptophis, Micrurus, Drymarchon, Drymobius, Bothrops,* and this individual. Presumably all came from cultivated areas formerly Tropical Deciduous and Semi-Evergreen Forest in the immediate vicinity of the village. Although several specimens were collected at the foot of the Sierra Madre west of Gómez Farías, elevation 300 m., Sr. Cordoba reported that none came from the mountainside above this elevation.

T. fasciata is quite rare in museums. None was reported by either H. M. Smith or E. H. Taylor in their extensive Mexican collections. The following data are thus noteworthy: the specimen is a male, scale rows 17 throughout, ventrals number about 173 (body in two pieces); caudals 80; upper labials 7-8, lower labials 8-8; two pre- and two post-oculars; white (?yellow) bands number one on the nape, 19 on the body, and 16 on the tail, all are asymmetric except the anterior two and the two on either side of the vent. Over the anus the scales are very slightly keeled, and the body slightly compressed.

The present specimen gives credence to three old Veracruz records for *T. fasciata* questioned by Smith and Taylor (1945), and comprises a noteworthy range extension from central Veracruz.

Tropidodipsas sartorii — Gómez Farías and vicinity, *ca.* 350 m. (2); Rancho del Cielo, 1050 m. (3), MMNH (15-6-54); sawmills of Rancho del Cielo area, exact locality and elevation unknown (11); Rancho Viejo, 1200 m. (2); Aserradero de Refugio No. 2, 1680 m.; total, 20 specimens.

While the majority of the specimens came from the Cloud Forest, two from Gómez Farías indicate occurrence below Cloud Forest in foothill tropical habitats. The specimen from Aserradero de Refugio No. 2 was found under a log in a transition area between upper Cloud Forest and Humid Pine-Oak Forest. This individual and at least three others for which I have reliable data were collected in clearings rather than in the forest proper. One from Gómez Farías had been swallowed by a *Micrurus*.

Stomach contents of four specimens were comprised of snails, including the genus *Humboldtiana* in UMMZ 110993 (identified by G. A. Solem). Three females (191372, 110989, 110996) with enlarged oviducal eggs, ready to lay, were all collected in May.

Tropidodipsas is a widely ranging tropical genus which reaches its known northern limit in southwestern Tamaulipas. Two subspecies of *T. sartorii* currently are recognized in Mexico. Apparently all the Rancho del Cielo specimens had yellow body rings like those of *T. s. annulatus* of the Pacific slopes of Chiapas, rather than the orange or red rings of *T. s. sartorii* from Veracruz and Yucatán. *T. s. annulatus* is also characterized by its very regular body rings, complete about the body and tail (four specimens); in *sartorii* at least some bands are usually incomplete on the ventral surface (Smith and Taylor, 1945:150). In the Tamaulipan series nine have complete bands, 11 have at least one incomplete body band, and, in three, numerous bands are incomplete with one or more not touching the edge of the ventral scales. Further morphological study of these northern *sartorii* is warranted. As Smith and Darling (1952) suggested, they may represent a distinguishable race.

Storeria dekayi. — Darling and Smith (1954) listed *S. d. temporalineata* from 3 km. N of Limón in the Gómez Farías region, elevation *ca.* 80 m. and William E. Duellman has taken one 10 km. N of the same locality. I have not found this species.

Storeria occipitomaculata. — Lower edge of Pine-Oak Forest on trail from Rancho del Cielo to Lagua Zarca, 1450 m.; Agua Linda, 1800 m. (2); sawmills of the Rancho del Cielo vicinity, elevation unknown, (2); total, 5 specimens. Doubt concerning the exact source of the latter records prevents including *Storeria* in the Cloud Forest fauna. To date it is known only from the Humid Pine-Oak Forest. Trapido (1944) lists the altitudinal range of other Mexican specimens as 6000 to 8000 feet (1800 to 2400 m.). The localities are all in areas of humid montane forest and the distribution pattern of these records fits that of the "Northeast Madrean" group.

Trapido (1944) has pointed out the similarity between Mexican and eastern U. S. populations of *S. occipitomaculata* and based his recognition of *S. o. hidalgoensis* on certain features of color pattern, including belly color, and a higher number of ventrals. Of the Tamaulipan specimens at least two possessed red bellies in life (UMMZ 110999-111000) while a third, 102986, was faintly pinkish. Ventral plus caudal counts of five specimens are as follows: male 126 + 57, 127 + 51; females 132 + 47, 130 + 48, 127 + 50. These values fall within the range of *S. o. hidalgoensis.*

While disagreement as to the taxonomic status of the Mexican red-bellied snake exists (Smith and Taylor, 1945 and Taylor, 1942, considered it a distinct species), no one would deny the close resemblance between Mexican and eastern North American populations. The paleoecological problem is to determine when and under what circumstances this forest snake, presumably one that could find no congenial habitat on the Great Plains (Trapido, 1944:19), ranged across the now arid Río Grande embayment. Trapido (p. 28) suggested evolution out of Mexico. However, the direction of dispersal, whether out of or into Mexico, is not pertinent to this problem.

Thamnophis cyrtopsis. — Lagua Zarca and vicinity, 1500-1800 m. (3); La Joya de Salas, 1500 m. (2). These five specimens reveal local range in both Humid Pine-Oak Forest and drier Oak-Pine Woodland. *T. cyrtopsis* shares the former habitat with *T. mendax*.

The present range of *T. cyrtopsis* is largely in the pine-oak belt of the Mexican Plateau with occasional records from certain drier habitats (Oak Savanna, Piñon-Juniper, and Oak Chaparral ?) exclusive of the desert basins. Milstead (1953:376) wrote: "*Thamnophis cyrtopsis* is restricted to plateaus and mountains. The plains and broad valleys which surround and separate these highlands constitute ecological barriers to the species." In this case the distribution pattern of *T. c. cyrtopsis* in the forests and woodland of western and eastern Sierra Madres, and in some of the isolated ranges of the now arid basin and range province of the Central Plateau, probably represents post Wisconsin isolation and restriction of a once widely ranging subspecies. For this reason I consider the present pattern evidence of a former Trans-Plateau distribution.

In his recent revision Milstead (1953) examined two of the above snakes and considered them intergrades between *T. c. cyrtopsis* and *T. c. cyclides*.

Thomnophis marcianus. — "15 miles north of Ciudad Mante (Villa Juarez), Tamaulipas" (Shannon and Smith, 1949:501). No habitat data accompany this record.

Thamnophis mendax. — Rancho del Cielo and vicinity, 1050 m. (3); Valle de la Gruta, *ca.* 3 km. W of Rancho del Cielo, 1500 m. (2); below Lagua Zarca in Cloud Forest (?), *ca.* 1200 m.; Lagua Zarca, and immediate vicinity, 1500-1620 m. (2); W of Lagua Zarca on trail to La Joya de Salas, *ca.* 1800-2100 m. (4); total, 12 specimens.

Both Humid Pine-Oak Forest and the upper Cloud Forest at Valle de la Gruta and Lagua Zarca are included in the local range of this garter snake. There is only one reliable record for the species at the elevation of Rancho del Cielo (101207) in lower Cloud Forest. Along the trail to La Joya de Salas specimens were taken on the west slopes of the Sierra Madre, but none below the level of Humid Pine-Oak Forest.

A specimen captured by Darnell contained a large salamander, *Pseudoeurycea belli*; another contained *P. scandens*.

In his recent description of *T. mendax*, Walker (1955*c*) discussed relationships and distribution. Although relatives inhabit humid forests to the south in San Luis Potosí, this species is known only from the mountains above Gómez Farías.

Thamnophis sauritus. — *ca.* 7 km. W of Limón, along an irrigation canal, 100 m. (2); *ca.* 7 km. E of Chamal, 120 m. In southern Tamaulipas outside of the Gómez Farías region the ribbon snake was collected at about 900 m. elevation in both the Sierra de Tamaulipas and the Sierra Madre at Laguna Escondida southeast of Tula. Probably it is most numerous in the irrigation networks of the Río Frío and Río Boquilla.

Natrix rhombifera. — *ca.* 8 km. WSW of Limón, 120 m. (7). These were collected along an irrigation canal of the Ciudad Mante sugar-cane district.

Micrurus fulvius. — Pano Ayuctle, 100 m. (5); Chamal and vicinity, 150 m. (2), TU 15526; Gómez Farías, 350 m. (3); *ca.* 2 km. NW of Gómez Farías, above 500 m.; trail between Pano Ayuctle and Rancho del Cielo, 510 m.; road between Rancho del Cielo and Gómez Farías, *ca.* 750 m.; total, 14 specimens. *M. fulvius* reaches 900 m. in oak savanna in the Sierra de Tamaulipas.

I have sought unsuccessfully to obtain a Cloud Forest record for this species, or evidence for distributional overlap with the coral snake "mimic" of the Cloud Forest, *Pliocercus elapoides.* The above localities represent Tropical Deciduous and Semi-Evergreen Forest. In color pattern *Pliocercus* is remarkably difficult to distinguish from *Micrurus.* Both have a similar sequence of complete black, yellow, and red rings, the latter with black-tipped scales. A very minor color difference is the absence of a black chin in *Pliocercus.* In external morphology they are quite distinct, *Pliocercus* having a very long tail and 17 scale rows, *Micrurus* a short tail and 15 scale rows. Evidently the two do not occur together in the Gómez Farías region, and thus it is difficult to imagine how Batesian mimicry could account for the color identity.

Agkistrodon bilineatus. — Chamal, 2 km. N, 160 m. (3); Santa Inéz, 7 km. S on the Mexico City - Laredo highway, *ca.* 120 m. (Smith and Darling, 1952); total, 4 specimens.

The three from near Chamal were removed from a den by laborers in the process of clearing palm forest on the M. Nichols ranch. In their combined experience, Mr. Nichols and R. Derr, lifelong Chamal residents and farmers, had seen only one other snake of this type. The vernacular name for *A. bilineatus* used elsewhere, "cantil," is unknown in the Gómez Farías area. The name "metapil" occasionally used by Chamal residents may possibly refer to this species.

The head pattern of these snakes agrees with that described for *A. b. taylori* (Burger and Robertson, 1951), and they are probably referable to that subspecies.

Bothrops atrox. — Pano Ayuctle, 100 m. (3), TU 15681; Gómez Farías, 350 m. (4); Aserradero del Paraíso, 490 m.; trail between Pano Ayuctle and Rancho del Cielo, 750 m., B. E. Harrell 33; Rancho del Cielo and vicinity, (2); north of El Tigre, 1000 m.; total, 13 specimens.

Humid tropical forest including parts of the Tropical Deciduous Forest, Tropical Evergreen Forest, and lower sections of Cloud Forest are inhabited by the "cuatro narices." *Bothrops* reaches the northern limit of its range along the slopes of the Sierra Madre Oriental in the Gómez Farías region. It is known in Tropical Deciduous Forest only near the Sierra

Madre, and should be sought in lower parts of the Sierra de Tamaulipas. At Pano Ayuctle *Bothrops* is uncommon; one was captured after it swam across the Río Sabinas.

Mexican specimens are currently referred to the subspecies *B. a. asper.*

Crotalus durissus. — *ca.* 4 km. NW Gómez Farías above Aguacates, 700 m.; Rancho del Cielo and vicinity, 1050-1100 m. (4); Rancho Viejo, 1200 m.; above Agua de los Indios, 4 km. SSW of Rancho del Cielo, *ca.* 1300 m.; Valle de la Gruta, 3 km. W of Rancho del Cielo, 1500 m.; 1 km. SE of La Gloria, 1500 m.; La Joya de Salas, above 1550 m. (2); total, 11 specimens.

It is remarkable that the tropical rattlesnake, primarily a savanna inhabitant through most of its extensive range in the lowlands of South and Central America, should occur at high elevations in upper Cloud Forest at the extreme northern limit of its range. In addition to Cloud Forest, *C. durissus* was collected in open Oak-Pine Woodland near La Joya de Salas, where *C. molossus* might be expected instead, and was seen as high as 1680 m. in Pine-Oak Forest near Aserradero del Refugio No. 2. In the lowlands at Pano Ayuctle *Crotalus* is allegedly rare. Sr. Marzo Dueños, a lifelong resident, could recall only one rattlesnake in that area, found during the hurricane and flood of late summer in 1951. This record could represent either *C. atrox* or *C. durissus.* Undoubtedly *C. atrox* occurs in the Gómez Farías region; C. M. Bogert reported (in litt.) a badly smashed specimen from north of Llera.

Stomach contents of a Cloud Forest *C. durissus* included a cave rat, *Neotoma*; a specimen from La Joya de Salas had devoured a tree squirrel, *Sciurus alleni.*

The present series was identified by L. M. Klauber as *C. d. totonacus.*

Crotalus lepidus. — Vicinity of Rancho del Cielo, elevation unknown, (3); Rancho Viejo, 1200 m. (2); Agua de los Indios, 4 km. SSW of Rancho del Cielo, 1300 m. (2); Lagua Zarca, 1500-1590 m. (2); western side of the Sierra Madre, *ca.* 5 km. SE of La Joya de Salas, *ca.* 1950 m. (2); Agua Linda, 1800 m. (4); total, 15 specimens.

Humid Pine-Oak Forest, the typical habitat of *C. l. morulus* in Tamaulipas, is more mesic than areas generally inhabited by *C. lepidus* populations. Although the upper parts of Cloud Forest are also inhabited, the presence of this species is not definitely established in lower Cloud Forest at Rancho del Cielo. Nos. 110926-8 were all purchased from sawmill workers at a time when a lumber road was being constructed to Valle de la Gruta, and they may all have come from above 1200 m. Four specimens collected at Agua Linda were found under logs in a small clearing surrounded by heavy forest.

A part of this series formed the type material of the subspecies *C. l. morulus* (Klauber, 1952). In addition to 15 specimens now known from the Gómez Farías region, one was collected in Humid Pine-Oak Forest near Chihue (1860 m.) northwest of Ciudad Victoria.

Order Testudines, Turtles

Terrapene mexicana. — Pano Ayuctle, 100 m. (5); Gómez Farías, 300 m.; total, 6 specimens.

In dry lowland habitats including Tropical Deciduous Forest, Thorn Forest, and possibly Thorn Scrub, the Mexican box turtle may be found. Humid tropical habitats including Rainforest of southern Veracruz probably serve to isolate populations of *Terrapene* in the northern tip of the Yucatán Peninsula and in northeastern Mexico. This distribution is paralleled by several other species, including *Laemanctus serratus*, *Hypopachus cuneus*, and *Sceloporus serrifer* (see p. 93). *T. mexicana* is a close relative of *T. carolina* and should possibly be considered conspecific with the latter.

BIOGEOGRAPHIC ANALYSIS

In preceding sections I have described the major environmental types of the Gómez Farías region, comparing them with similar environments elsewhere in northeastern Mexico. The herpetological fauna was described in terms of these environmental types. Ninety-four species representing five salamanders, 20 frogs, 24 lizards, 44 snakes, and one turtle have been found in the Gómez Farías region, a sample that I estimate represents at least 75 per cent of the total terrestrial fauna. Among these are 36 species of tropical distribution at or near their known northern ranges in eastern Mexico. In addition at least five interior temperate North American species find their southern limits in southwestern Tamaulipas. With this basis it is possible to explore the ecological and historical questions introduced earlier.

Zonation

Adequate materials for a zonal analysis are not assembled rapidly. Ideally they include a transect across a variety of climatic and vegetation types. Interference by man should be minimal. The fauna should be well known and the relative abundance as well as the altitudinal limits of each species understood. Systematic collecting in the manner of Hairston (1949) at regular altitudinal intervals on all slopes is desirable.

In the Gómez Farías region these conditions were only partly met. Relative abundance of the various species is largely unknown and the altitudinal limits of only the more common species can be approximated. Nevertheless, certain distribution patterns are evident. In selecting species for zonal patterns I have excluded as too poorly known all of those recorded from fewer than five localities. To date these number 45, or about half the fauna. Only 37 species are considered known with sufficient confidence to justify mapping their zonal range.

Rana pipiens ranges through the entire area wherever suitable breeding ponds appear. As the only member of the herpetological fauna found in all eight major vegetation types, this species is of interest.

Viewing the herpetological fauna as a whole, three major zonal groups can be identified (Fig. 7); (1) a humid montane forest group; (2) a thorn forest, dry oak-pine and tropical-deciduous forest group of both the lowlands and the dry interior, and (3) a subhumid evergreen or deciduous

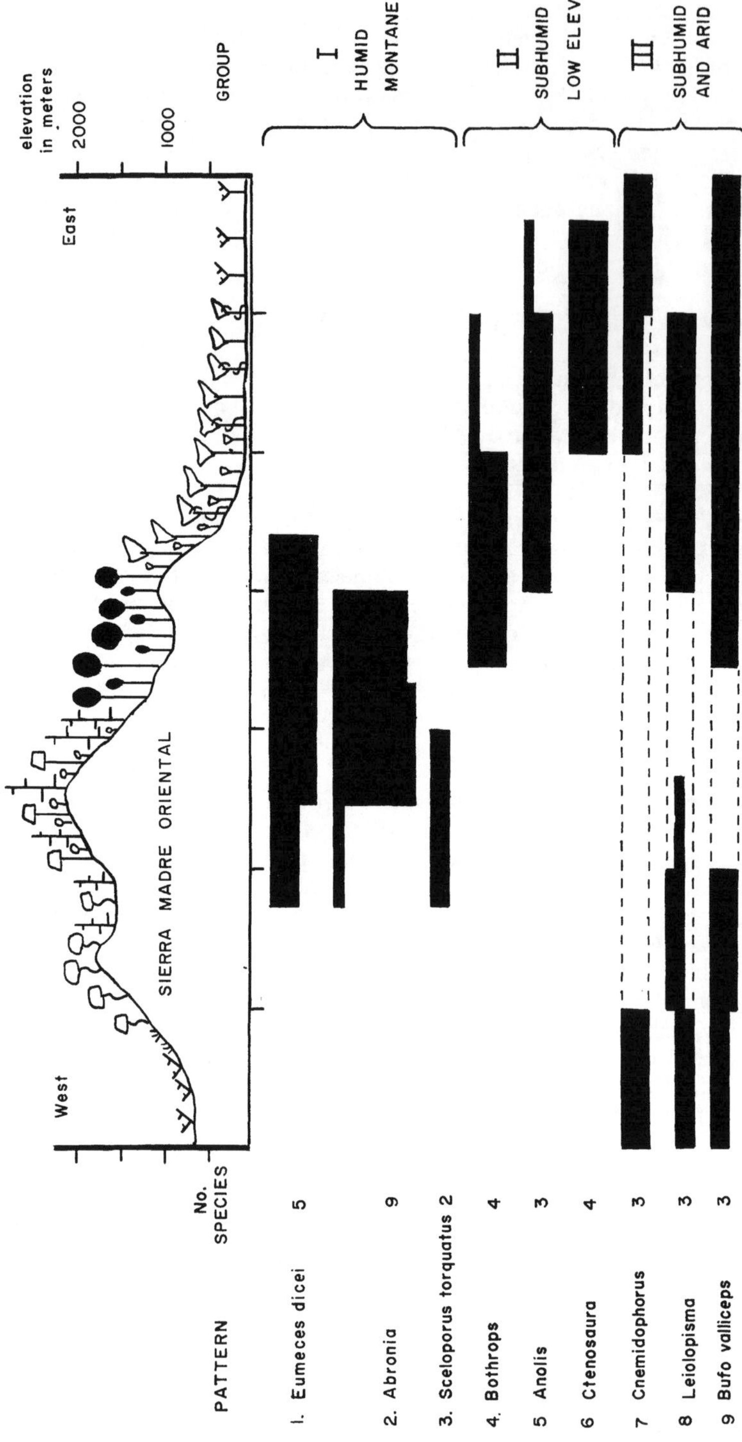

Fig. 7. Zonal distribution of reptiles and amphibians in the Gómez Farías region, based on 37 species recorded from five or more localities. Seven vegetation types figured are given equivalent areas. For key to vegetation symbols see p. 27, Fig. 6.

tropical lowland forest group which avoids the interior scrub. These groups are subdivided into nine patterns of distribution, each identified by the name of a typical member. A few of the thorn forest and thorn desert records illustrated in Figure 7 are postulated on the basis of information obtained from immediately outside the Gómez Farías region. Lack of faunal information requires omission of the Chaparral zone.

Group I. — The humid montane forest, Pine-Oak and Cloud Forest, were studied more intensively than other formations in the Gómez Farías region. A number of species are confined to these forests. In addition to 16 common reptiles and amphibians, a number of birds such as *Trogon mexicanus, Lepidocolaptes affinis, Mitrephanes phaeocercus, Empidonax difficilis, Catharus occidentalis, Catharus mexicanus, Aphelocoma ultramarina,* and *Basileuterus belli* have a similar range. The mammals are poorly known, but *Reithrodontomys mexicanus* and *Cryptotis mexicana* may belong in this group. All of the species in Group I are ecologically isolated in the Gómez Farías region by arid interior valleys to the north and west or by lowland Tropical Deciduous Forest on the south and east (Map 2).

1. *Eumeces dicei* Pattern. Five species including *Crotalus durissus, Chiropterotriton multidentata, Eleutherodactylus hidalgoensis, Syrrhophus latodactylus,* and *Eumeces dicei* occur mainly in Humid Pine-Oak Forest and Cloud Forest, but they also descend to 400 m. in Tropical Evergreen Forest. Three of them, *Crotalus, Eumeces,* and *Eleutherodactylus,* range into Dry Oak-Pine Woodland near La Joya de Salas (1500 m.).

2. *Abronia* Pattern. This large, important group includes nine species, *Abronia taeniata, Pseudoeurycea belli, P. cephalica, P. scandens, Chiropterotriton chondrostega, Lepidophyma flavimaculatum, Rhadinaea crassa, Thamnophis mendax,* and *Crotalus lepidus.* The local distribution of these is quite similar to that of Pattern 1 except that none of them ranges below the Cloud Forest (1000 m.) and only one, *Rhadinaea,* was found in Dry Oak-Pine Woodland near La Joya. *Crotalus lepidus* is not definitely known from lower Cloud Forest. For the remaining species in this group the lower limit of Cloud Forest constitutes an important faunal boundary.

3. *Sceloporus torquatus* Pattern. *Sceloporus torquatus* and *S. grammicus* occur both in Dry Oak-Pine Woodland near La Joya de Salas and in Humid Pine-Oak Forest. Unlike other members of Group I, they avoid Cloud Forest. *Thamnophis cyrtopsis* may also belong in Pattern 3.

Group II. — A number of species at or very near their northern range limit occur only in the subhumid tropical forests of the east slopes of the mountains. While they may overlap broadly with members of Group III in this area, they are largely confined to, or find centers of abundance in, Tropical Evergreen Forest and Tropical Deciduous Forest. The number of these which also occur in Thorn Forest and Scrub on the outer coastal plain is unknown; presumably most of them do not. In Group II are most of the "typical" tropical forest species that reach their latitudinal range limit in the Gómez Farías region. Of the other vertebrates the birds *Nyctibius, Geotrygon, Claravis,* and mammals *Heterogeomys, Oryzomys fulvescens,* and *Centurio* have a similar local range.

4. *Bothrops* Pattern. Unlike other members of Group II, *Bothrops atrox, Dryadophis melanolomus, Syrrhophus cystignathoides*, and *Thalerophis mexicanus* share the distinction of ranging from adjacent Tropical Evergreen and Deciduous Forest into lower Cloud Forest to about 1200 m.

5. *Anolis* Pattern. At least three species including *Anolis sericeus, Micrurus fulvius*, and *Laemanctus serratus* occur in Tropical Semi-Evergreen and Deciduous Forest. They are unknown from Cloud Forest, and their upper altitudinal limit is reached at about 800 to 1000 m. *Micrurus* and *Anolis* may be expected to occur in Thorn Forest.

6. *Ctenosaura* Pattern. This assemblage favors drier tropical habitats than do other members of Group II. None of the forms included, *Ctenosaura acanthura, Hyla staufferi, Ameiva undulata*, and *Leptodeira maculata*, is known in Cloud Forest or Tropical Semi-Evergreen Forest. They range through Tropical Deciduous and possibly Thorn Forest below 600 m.

Group III. — In contrast to Group I, a number of species surround part or all of the humid forest area. A few enter the humid forests in clearings, along trails, or in the other sunny, dry, microhabitats. Additional collecting should add many more examples of this type. Since the arid and subhumid vegetation types through which these forms range are widespread and continuous in most of northeastern Mexico, these species are continuously distributed north of the Gómez Farías region.

In many respects the Thorn Scrub and Thorn Desert of the interior valley are similar to Thorn Forest and Scrub of the coastal plain. The Dry Oak-Pine Woodland is probably just as dry and only slightly cooler than the lowland and foothill Tropical Deciduous Forest. It is not surprising that these habitats have many species in common. Among the birds, the coppery-tailed trogon *(Trogon elegans)*, brown jay *(Psilorhinus morio)*, and blue grosbeak *(Guiraca caerulea)* may share a zonal range with the following reptiles and amphibians.

7. *Cnemidophorus* Pattern. *Sceloporus olivaceus, Masticophis flagellum*, and *Cnemidophorus sacki* occupy only the driest habitats at fairly low elevation (below 900 m.).

8. *Leiolopisma* Pattern. *Leiolopisma sylvicolum* (sp. ?), *Eleutherodactylus augusti*, and *Leptotyphlops myopicus* range into more humid tropical forests (Tropical Deciduous Forest) and occur at higher elevation (Dry Oak-Pine Woodlands 1500 m.) than the preceding. They may be absent from the driest parts of the coastal plain. Data on lowland range of all three is deficient. The altitudinal range is 100 to 1500 m., *E. latrans* reaching 2000 m.

9. *Bufo valliceps* Pattern. *Bufo valliceps, Sceloporus variabilis*, and *Leptodeira septentrionalis* differ from members of the preceding pattern in that they enter lower portions of the Cloud Forest. Here *Sceloporus* and *Leptodeira* seek out the sunny, drier microhabitats. Three additional species, *Smilisca baudinii, Drymobius margaritiferus*, and *Bufo marinus* have identical ranges on the east slopes of the Sierra Madre. To date they are unknown from the west slope, with the exception of a sight record of *Drymobius*; however, they may be expected there and will probably prove to be members of Pattern 9.

The nine patterns and three groups described above by no means represent all the zonal types that exist. Among the rarer species (fewer than five locality records), certain other patterns appear. *Phrynosoma cornutum* and *Bufo punctatus* are found only in the Jaumave Valley. Possibly they belong to an element confined to this arid interior basin. *Hyla eximia, Sceloporus scalaris,* and a number of others are known only from Dry Oak-Pine Woodland near La Joya de Salas. They may indicate a distinct arid or subhumid montane element. *Pliocercus elapoides* is unique in being fairly common in Cloud Forest but unknown outside this formation. In general my information on Chaparral, Dry Oak Woodland, and interior basin and valley Thorn Scrub is not sufficient to predict what patterns are involved west of the eastern slopes. Future zonal investigations in this area should focus on the west side of the Sierra de Guatemala.

Sceloporus (fence lizards, nine species) may serve to illustrate zonal segregation within a genus. No other genera are represented by more than four species, and 45 of the 63 genera present have only a single representative. Five species of large-sized *Sceloporus* (adults exceeding 85 mm. snout to vent) represent potential competitors. The only point at which two of them were found together is in the La Joya Basin where *S. torquatus* and *S. jarrovii* overlap slightly (Table VI). Four other *Sceloporus* of smaller size are also found in the La Joya basin. Presumably microenvironmental rather than zonal isolation separates them.

TABLE VI

Zonal Segregation of Five Large *Sceloporus* in the Gómez Farías Region.

Species	Altitudinal Range in Meters	Vegetation Type	Habit
S. olivaceus	100 to 1000	Thorn Scrub and Thorn Desert	Arboreal
S. serrifer	100 to 300	Tropical Deciduous Forest	Arboreal
S. cyanogenys	1050	Cloud Forest (clearing only)	Arboreal, saxicolous
S. torquatus	1400 to 2100	Humid Pine-Oak Forest, Dry Oak-Pine Woodland	Arboreal, saxicolous
S. jarrovii	1500 to 2100	Dry Oak-Pine Woodland, Chaparral	Saxicolous

Intra-zonal ecological isolation. The above zonal analysis describes a primary basis for segregation and explains only why certain species do not compete. Further inquiry into details of spatial isolation within each plant formation is necessary where more than one species of a genus occur together. This is true of the smaller *Sceloporus* (*parvus, scalaris, grammicus,* and *variabilis,* all in the La Joya basin), of the snake genus *Elaphe* (*flavirufa, triaspis,* and *guttata,* all in Tropical Deciduous Forest), and of the salamanders. The latter demonstrate a certain degree of habitat isolation.

Two genera comprising five species of plethodontid salamanders are found together in Humid Montane Forest. Within each genus a pair of species appears, one larger with relatively longer legs (*P. scandens, C. multidentata*), the other smaller with shorter legs (*P. cephalica, C. chondrostega*). The larger species is scansorial in habit, occupying cave walls and ceilings, and climbing into bromeliads. The smaller is mainly terrestrial. Although closely related, they are not known to interbreed. The number of specimens collected in various habitats is listed in Table VII.

TABLE VII

Species Pairs of Tamaulipan Salamanders in the Genera *Pseudoeurycea* and *Chiropterotriton*. Figures refer to number of specimens collected in each habitat. Only Cloud Forest and Humid Pine-Oak Forest collections are included.

Larger size, relatively longer legs.	Smaller, shorter legs.	Troglodytic	Arboreal	Terrestrial	Total
P. scandens		61	7	47	115
C. multidentata		14	69	2	85
	P. cephalica			21	21
	C. chondrostega		32	116	148
	Total	75	108	186	369

If we assume that the scansorial form is partly removed from competition with its terrestrial relative, the possibility of intergeneric competition between similarly adapted types (*Pseudoeurycea cephalica* and *Chiropterotriton chondrostega*) remains. I have not investigated this beyond noting that *Pseudoeurycea* is much larger than *Chiropterotriton.*

A fifth salamander in the montane forests, *P. belli*, is much larger, mainly terrestrial, and possibly more fossorial than the others.

Schmidt (1936) has described the zonal distribution of nine species of salamanders on Volcan Tajumulco in Guatemala. No more than three occur together at any single altitude and only in parts of the Cloud Forest (5000′-9700′) did he find more than two species.

Vegetation and the Border Tropical Fauna

From 18^0 to 26^0 N. lat. the climatic gradient along the tropical lowlands of eastern Mexico involves the following factors: (1) a sharp reduction in annual precipitation; (2) a shift from seasonal to nonseasonal distribution of rainfall; (3) a gradual increase in seasonal temperature range, and (4) a slight drop in mean annual temperature. The vegetation shifts from humid to arid tropical and finally to arid temperate as various formations reach their northern limits. A corresponding diminution of tropical elements would be expected in the fauna.

An analysis of the faunal gradient throughout this distance is beyond my present purpose. It is notable, however, that in this region 12 of the 23 families of birds in Mayr's "South American Element" (Mayr, 1946) reach their Atlantic lowland latitudinal limits. These are presented in Table VIII, together with the vegetation type with which they seem to be associated.

TABLE VIII

Latitudinal Limits of Some Tropical Bird Families and Associated Vegetation Types

Family	Approximate Latitudinal Limit		Possible Limiting Vegetation Type
Eurypigidae Galbulidae Bucconidae Pipridae	17° 30′ to 18°	South Veracruz and Tabasco	Rainforest
Ramphastidae Furnariidae Coerebidae	22°	SE San Luis Potosí	Tropical Evergreen Forest
Nyctibiidae Tinamidae Dendrocolaptidae Formicariidae	23° 30′	South Tamaulipas	Tropical Deciduous Forest
Cotingidae	26°	South Texas	Thorn Forest and Scrub

In addition it appears that the Xilitla region of southeastern San Luis Potosí, near the limit of continuously distributed Tropical Evergreen Forest (and Cloud Forest?), has a decidedly richer tropical fauna than the Gómez Farías region (Taylor, 1949-53; Davis, 1952; Lowery and Newman, 1951). That the presence of Tropical Evergreen Forest is entirely responsible for the richer San Luis Potosí fauna is doubtful; however, this formation does bring with it an animal environment not well developed north of eastern San Luis Potosí.

Much more study of tropical plant formations is needed before their influence on animal distribution can be firmly established. Present information on northeastern Mexico indicates that two formations, Tropical Deciduous Forest and Thorn Forest and Scrub, correspond closely to the local ranges of various animals. On the other hand despite their distinctive structural character, the small, isolated outposts of Tropical Evergreen Forest and Cloud Forest in Tamaulipas appear to play a minor role in determining the nature of the tropical fauna at this latitude.

Perhaps the most striking biogeographic feature of northeastern Mexico is the faunal "break" in southern Tamaulipas, corresponding to the northern limit of Tropical Deciduous Forest. It is this faunal line which drew the attention of Salvin and Godman and subsequent authors. Treating this area as the limit of the "Neotropical Region," however, obscures the nature of the Gulf Coast gradient, of which the northern limit of Tropical Deciduous Forest is but a single step.

Within the Gómez Farías region the following four faunal components are recognized as part of the Gulf gradient. Three concern the northward penetration of tropical faunas, the fourth is the converse, an interior temperate fauna extending from the north into arid border tropical habitats.

Dry Lowland Tropical. An important group of reptiles and amphibians reach their northern range limit in Thorn Scrub of southern Texas (*ca.* 800 mm. annual rainfall). In the Gómez Farías region some of these occur in both humid and arid habitats, but their tolerance for the latter permits them to penetrate farther north through the Thorn Forest and Scrub than do the other lowland species confined to wetter tropical environments. Species in this group include: *Rhinophrynus dorsalis*, *Bufo horribilis*, *B. valliceps*, *Leptodactylus labialis*, *Smilisca baudinii*, *Hypopachus cuneus*, *Sceloporus variabilis*, *Coniophanes imperialis*, *Drymobius margaritiferus*, *Drymarchon corais*, and *Leptodeira annulata*. All except *Rhinophrynus* reach southern Texas; *Bufo valliceps* extends into Louisiana; *Micrurus* and *Drymarchon* continue eastward to Florida.

Subhumid Lowland Tropical. The striking faunal boundary that corresponds with the northern limit of Tropical Deciduous Forest in southern Tamaulipas has induced many zoogeographers (see recent maps by Smith, 1949, and Tamayo, 1949), to consider this the line between "Nearctic" and "Neotropical" regions. A variety of animals follow lowland tropical habitats north to the limit of the Tropical Deciduous Forest, roughly the Tropic of Cancer. Some of these are continuous in distribution between Tamaulipas and Argentina *(Spilotes pullatus)*; others occur only in Middle America *(Laemanctus)*.

Common species in this group found both in the Tamaulipan lowlands, foothills, and occasionally Cloud Forest include: *Syrrhophus cystignathoides*, *Laemanctus serratus*, *Anolis sericeus*, *Dryadophis melanolomus*, *Elaphe flavirufa*, *E. triaspis*, *Leptophis mexicanus*, and *Bothrops atrox*. Those in the lowlands only are: *Phrynohyas spilomma*, *Hyla staufferi*, *Leptodactylus melanonotus*, *Ctenosaura acanthura*, *Sceloporus serrifer*, *Ameiva undulata*, *Constrictor constrictor*, *Imantodes cenchoa*, and *Leptodeira maculata*.

Humid Tropical. A very small but interesting group, elsewhere found mainly in lowland Tropical Evergreen Forest and Rainforest, reaches its northern limit in the Gómez Farías region or shortly to the north. Here they occur primarily in Cloud Forest, the nearest equivalent in humidity to the Rainforest farther south. *Lepidophyma flavimaculatum*, *Tropidodipsas sartorii*, and *Pliocercus elapoides* share this range. Surprisingly, all appear in the isolated Coffee Belt (Cloud Forest and Tropical Evergreen Forest) on the Pacific slopes of Chiapas and Guatemala. None is known from Pacific slopes north of the Balsas River.

Arid Interior. In addition to northward decline in tropical species the Gulf lowland environmental gradient can be measured conversely in terms of southward decline of an arid interior temperate fauna. This penetrates eastern Mexico from the north and west, a few species reaching central Veracruz. Most of the Arid Interior species do not extend south along the coastal plain beyond southern Tamaulipas, eastern San Luis Potosí, and northern Veracruz. Here the habitats they favor interdigitate with and give way to more humid tropical forests. Along the Plateau escarpment many of them follow arid valleys and rainshadow slopes to somewhat lower latitudes than they reach on the coastal plain. None of these species reaches outer Yucatán where arid habitats similar to those in southern Tamaulipas are present. Of the Gómez Farías fauna the following species belong to the "Arid Interior" component: *Bufo punctatus*, *Scaphiopus couchi*, *Rhinoceilus lecontei*, *Holbrookia texana*, *Phrynosoma cornutum*, *Sceloporus olivaceus*, and *Masticophis taeniatus*. Others in central Tamaulipas, not yet found in the Gómez Farías region, include *Crotalus atrox*, *Hypsiglena* sp., *Gopherus berlandieri*, and *Arizona elegans*. The Southwestern Desert Fauna, which on ecological grounds might be expected in a few areas (such as the Jaumave Valley), is represented mainly by aggressive, wide-ranging desert forms such as *Arizona*, *Hypsiglena*, and *Rhinoceilus*. In Tamaulipas the arid coastal plain fauna shows no marked degree of endemism.

Pleistocene Dispersal Routes in Eastern Mexico

In describing lowland faunal distribution I have indicated a correlation with the climatic and vegetational gradients in northeastern Mexico. On the assumption that the present arrangement has been of relatively short duration and that Pleistocene climatic influence has intervened, we may seek faunal evidence of former distribution patterns. In addition to isolated habitats disjunct distribution of various animals presumably represents the outcome of Pleistocene climatic change. There is evidence of three former dispersal routes which are no longer continuous. The present relict faunas are identified as the Northeast Madrean, Trans-Plateau, and Gulf Arc components. The floristic relationship between humid montane forests in eastern Mexico and those in southeastern United States, treated elsewhere (Martin and Harrell, 1957) is considered largely the outcome of pre-Pleistocene events. There is no marked faunal evidence of a direct Pleistocene connection between forest faunas in the mountains of eastern Mexico and those in eastern United States.

Regarding disjunct animal distributions within Mexico certain difficulties in the nature of the evidence should be noted. First, the apparent break in range of some species may not be real. The steady flow of range extensions within various Mexican states reflects a growing but still sketchy outline of animal distribution. The best negative evidence may come from those animals so closely tied to a specific habitat (Plethodontid salamanders and humid forest) that gaps in present ranges can be assumed on an ecological rather than an observational basis.

A second problem concerns the impact of both pre- and post-Columbian cultural activity. At the arrival of Cortez the population of central Mexico is estimated at eleven million (Cook and Simpson, 1948). The Spanish invasion and the introduction of domestic animals, notably cattle, horses, asses, sheep, and goats, and the destruction of forest for mine timbers and fuel created a new level of cultural disturbance (Simpson, 1952). Many existing distributional gaps may be the result of cultural rather than climatic change.

A third problem involves the mainfold effect of Pleistocene climatic change. Following four major ice advances, three major retreats, and unknown lesser oscillations before and during this sequence, the present faunal arrangement must represent the summation of a complex climatic history. The precise time at which any particular separation occurred, and the subsequent history of isolation and secondary contact (if any) is not readily deduced from existing ranges. For these reasons the present use of the term Pleistocene is noncommital with regard to chronological detail.

A fourth reservation concerns adequacy of taxonomic information. Obviously disjunct distribution patterns based on species or genera which have undergone thorough revision (*Reithrodontomys*, Hooper, 1952) deserve more weight than those less thoroughly treated (*Terrapene*).

With these reservations in mind I consider the following faunal components as evidence of Pleistocene climatic change.

Northeast Madrean component (Map 3). In humid montane forests of the Gómez Farías region are found at least 11 species which do not, or only rarely, range below 1000 m. and occur above this elevation only in mesic forests, either Cloud Forest or Pine-Oak Forest. Members of this group appear in the mountains of Nuevo León, Hidalgo, San Luis Potosí, Puebla, and northern Veracruz (Map 3). I have mapped the distribution of eight species or vicariant representatives found both in the Gómez Farías region and in Hidalgo. They include: *Pseudoeurycea cephalica rubrimembris* and *P. c. manni*; *Chiropterotriton chondrostega*; the species group comprising *C. multidentata*, *C. arborea*, and *C. mosaueri*; *Pseudoeurycea scandens*; *Eleutherodactylus hidalgoensis*; *Geophis semiannulatus*; *Rhadinaea crassa* and its close relative *R. gaigeae;* and *Storeria occipitomaculata hidalgoensis*. *Abronia taeniata* belongs in this group but its distribution was not mapped. In addition, *Pseudoeurycea belli* and *Sceloporus jarrovii immucronatus* follow this pattern through part of their range.

As is evident from the vegetation map of the Gómez Farías region (Map 2) the humid montane forests which these species occupy in the Sierra de Guatemala are entirely surrounded by much drier habitat. The forests east and south of the mountains are exclusively lowland tropical. To the west occur habitats equally unsuitable. A series of ridges and valleys separate the Sierra de Guatemala from the Mexican Plateau proper. The first of the valleys, extending from Ocampo to Jaumave, reaches its highest point (and the highest point connecting the Sierra de Guatemala with adjacent ranges) west of La Joya de Salas at 1000 m. This entire valley is covered with Thorn Forest.

To the north continuous humid montane forest can be traced no farther along the Sierra de Guatemala than the Río Guayalejo gap where another physiographic break intervenes and a semidesert barrier is encountered.

Outside the Gómez Farías region many other isolated montane forests are known, mainly Pine-Oak Forest but including Cloud Forest near Xilitla and in Hidalgo. The localities shown on Map 3 represent forests in which Northeast Madrean species have been reported. The close taxonomic affinity of isolated populations comprising the same or vicariant species in the Northeast Madrean group implies relatively recent (although not necessarily post-Sangamon) dispersal across the dry San Luis Potosian gap and other smaller ecological barriers separating these populations today. The fact that all species under consideration are terrestrial vertebrates with limited dispersal powers and, so far as known, with a very specific confinement to humid montane habitats, virtually eliminates the possibility of accidental arrival in these habitat islands. A synopsis of conditions south of Tamaulipas may further illustrate the nature of the present barrier.

Between southwestern Tamaulipas and northern Hidalgo the Plateau

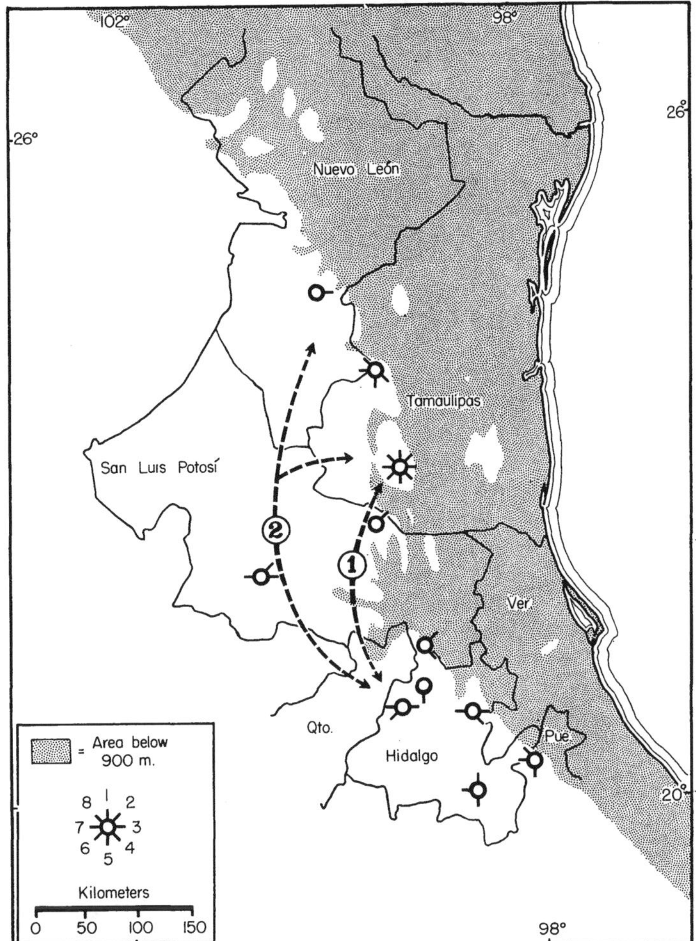

Map 3. Distribution of eight species or species groups in the northeast Madrean component. Numbers on the spokes of each circle (see inset) refer to the following:

1. *Geophis semiannulatus*
2. *Rhadinaea crassa* and *R. gaigeae*
3. *Storeria occipitomaculata hidalgoensis*
4. *Eleutherodactylus hidalgoensis*
5. *Pseudoeurycea cephalica* (part)
6. *Chiropterotriton chondrostega*
7. *C. multidentata, C. arborea,* and *C. mosaueri*
8. *Pseudoeurycea scandens*

The following humid montane localities are shown: Nuevo León, hills above Pablillo; Tamaulipas, Chihue region and Gómez Farías region; San Luis Potosí, Xilitla, El Platanito, and Alvarez; Hidalgo, Jacala region, Durango region, El Chico-Guerrero region, and Zacualtipan-Tianguistengo region; Puebla, Nexaca.

These humid montane habitats are presently isolated by arid basins and valleys. There are two possible routes along which they may have been interconnected by a Pleistocene forest corridor. Route 1 follows the edge of the escarpment; route 2 follows the higher interior ranges.

escarpment is low, not exceeding 1600 m. in San Luis Potosí and cresting below this elevation at most points. Two major drainage systems, the Río Moctezuma and the Río Santa María dissect the escarpment and divide the rather dry, oak-sweet gum forests (decidedly drier than Tamaulipan Cloud Forest) that lie on the east slope of the escarpment. Dalquest (1953:9), found that the Río Santa María forms a faunal corridor along which lowland tropical species move into the Río Verde valley "... without encountering the oak belt along the crest of the Sierra Madre." From Dalquest's description it appears that the Río Santa María gorge constitutes as much of a barrier to montane forest species in San Luis Potosí as does the Río Guayalejo in Tamaulipas.

Behind the escarpment in southern San Luis Potosí lies the arid valley of the Río Verde. The town of Río Verde, elevation 990 m., receives 538 mm. annual precipitation, and such desert plants as *Larrea* typify the landscape (Goldman, 1951:246). If the Río Verde valley and the Río Santa María gorge draining it comprised an insurmountable obstacle to past dispersal of the Northeast Madrean forest faunas, an alternate connective route may have been involved. Isolated ranges between Amoles, Queretaro, Alvarez and Matehuala, San Luis Potosí, and Miquihuana, Tamaulipas (route 2 on Map 3), skirt the Río Verde basin. The presence of *Rhadinaea gaigeae*, *Storeria occipitomaculata*, and *Chiropterotriton multidentata* near Alvarez and the description of a forest of magnolia and pines 38 km. southwest of Río Verde (notes by Fugler in Taylor, 1953) indicate a Northeast Madrean biota.

How and under what climatic conditions were the forest animals in the Northeast Madrean group connected? Muller (1939) estimates precipitation in montane mesic forest of Nuevo León at 1200-1800 mm.; Cloud Forest in Tamaulipas probably receives less than 2000 mm. only in the driest of years and the Humid Pine-Oak forests probably receive an equivalent amount. The rainfall in the dry basins and valleys isolating the montane forests can be estimated from the list of localities presented in Table IX.

These figures (from Vivo y Gomez, 1946) characterize the precipitation in or near the bottom of the basins in question; the basin divides probably receive slightly more, perhaps 600 to 800 mm. If we accept Muller's figure of 1200 mm. as the minimum to support Montane Mesic Forest, the former continuity of this habitat would require a doubling of present rainfall.

Trans-Plateau component. In Dry Oak-Pine Forest and woodland, which surround but do not enclose the treeless Central Plateau on three sides, a distinctive faunal group is found. In the Gómez Farías region *Hyla eximia*, *Sceloporus grammicus*, *S. torquatus*, *S. scalaris*, *Thamnophis cyrtopsis*, and possibly *Crotalus lepidus* and *Eleutherodactylus augusti*, represent oak-woodland and dry-forest species. All might be assumed to have arrived in northeastern Mexico from the south. It is also possible, however, that some of them arrived from the west in late Pleistocene time, across an oak savanna now eliminated in the arid interior plateau. The distribution of *Sceloporus scalaris slevini*, *S. grammicus disparilis*, *Thamnophis c. cyrtopsis*, and *Barisia imbricata ciliaris* suggests such a

TABLE IX

Annual Precipitation of Intermontane Basins and Valleys in Northeastern Mexico

Locality	Degrees N. lat.	Degrees W. long.	Elevation	Mean Annual Precipitation in mm.
Nuevo León:				
Galeana	24° 50′	100° 04′	1654	473
Tamaulipas:				
Jaumave	23° 25′	99° 19′	735	568
San Luis Potosí:				
Matehuala	23° 39′	100° 38′	1651	422
Charcas	23° 08′	101° 07′	2075	387
Cerritos	22° 26′	100° 17′	1153	549
San Luis Potosí	22° 09′	101° 05′	1877	361
Río Verde	21° 56′	100°	991	538
San Ciro	21° 38′	99° 50′	883	753

history. Each of these races, which are largely confined to montane woodland or savannas, is represented in both the Sierra Madre Oriental and Sierra Madre Occidental. To the south they are replaced and thus isolated, by related subspecies in the transverse volcanic district of Central Mexico. In like fashion the completely collared *Sceloporus torquatus* of Durango may have been derived from eastern rather than western *torquatus*. *Peromyscus megalotis amoles* in Queretaro which Hooper (1952) found most similar to the section of the Sierra Madre Occidental subspecies, *P. m. zacatecae*, from Jalisco and Michoacán, may represent a more southerly Trans-Plateau relict. In describing a large land snail, *Humboldtiana durangoensis*, of pine-oak woodland, Solem (1954) pointed out that on conchological characters it is nearest to *H. taylori* and *H. nuevoleonsis* of the Sierra Madre Oriental.

As morphological similarity at the subspecific level could represent parallel evolution rather than past connection between montane forms in eastern and western Mexico, other evidence is desirable. Outstanding in this regard is a chipmunk, *Eutamias bulleri*, which presently inhabits pine-oak forests and savannas of the Sierra Madre Occidental. Chipmunks are unknown from the forests of the Transverse Volcanic District and from most of the Sierra Madre Oriental, a significant piece of negative evidence in the case of a fairly conspicuous diurnal mammal. There are, however, two records on the eastern side of the Plateau in southeastern Coahuila (Howell, 1929, Baker, 1956). East of San Antonio de las Alazanas in the Sierra Madre Oriental, *E. bulleri* inhabits "... stands of pine, fir, and aspen at elevations no lower than 9000 feet" (Baker, 1956:210).

Baker adds *Sciurus alleni*, *Peromyscus truei*, and *Neotoma mexicana navus* to the montane mammals with closest affinities to the west Madrean fauna. Dispersal along a Trans-Plateau route (Map 4) similar to the route I propose is clearly mapped. Of parenthetical interest is Baker's

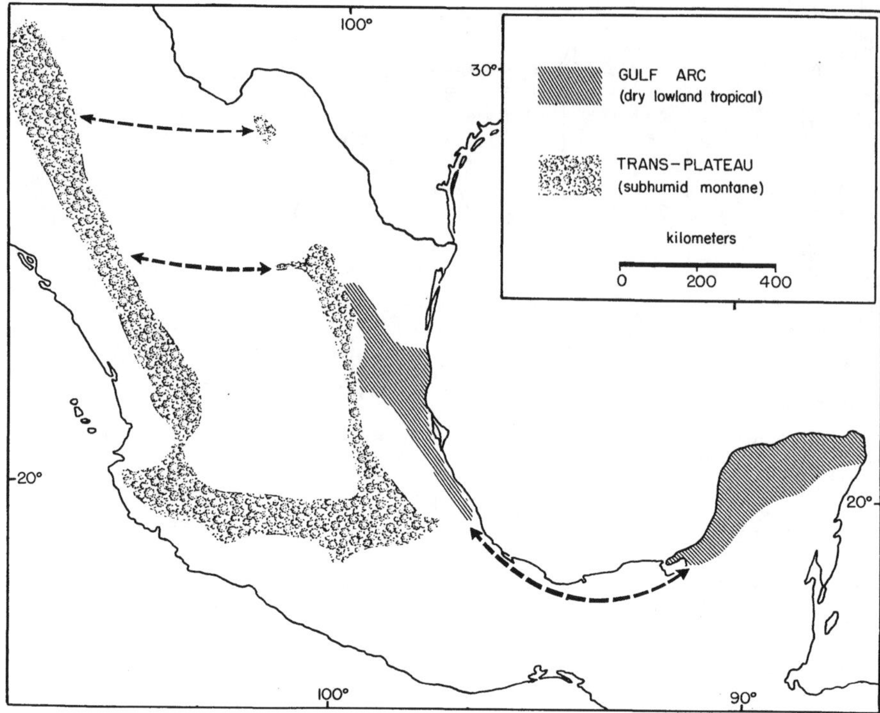

Map 4. Diagrammatic distribution of species in the Gulf Arc and Trans-Plateau components. Broken lines indicate possible Pleistocene dispersal routes presently closed by ecological barriers.

observation that all isolated montane Coahuilan mammals inhabiting ranges north of 26^0 are derived from west Texan and Rocky Mountain species rather than Madrean faunas.

Confirmation of the Trans-Plateau route should also be sought among the dominant oaks and pines that must have formed the past habitat corridor. Disjunct distribution of populations of *Pinus flexilis* and *P. arizonica* (Martinez, 1948) in eastern and western Mexico appear to support the zoological evidence.

Finally, there is a geological indication of moister conditions in the Central Plateau along latitude 25^0 N. during the Pleistocene. The paleoecological interpretation of elephant remains reported by Arellano (1951) from near Parras, Coahuila, includes the presence of permanent water, scattered trees, and tall grasses. Tentatively, Arellano dates this deposit as Kansan. He proposes that the maximum development of the bolson lake in the structural trough between Torreón and Saltillo was Pliocene. In tufa beds in the Sierra de Parras, Imlay (1936) discovered numerous molds and calcified replacements of coarse grasses, which may have grown in a marshy plain near one of the now extinct pluvial lakes. Rapid recent erosion in the Sierra de Parras is indicated by hanging canyons.

The isolated ranges topped with Oak-Pine Forest in Coahuila (Muller, 1947) appear to represent relicts of Trans-Plateau dispersal. Some magnitude of the climatic change necessary to achieve such a connection can be appreciated by examining meteorological records from western Coahuila and eastern Durango. Matamoros, Coahuila, elev. 1120 m., receives 238 mm. annual precipitation.

Additional taxonomic study of montane species is needed to understand the nature of Trans-Plateau dispersal. However, such an interpretation finds strong biogeographic support.

Gulf Arc component. Pleistocene climatic change is readily inferred from the relict distribution of montane forest and woodland faunas. By analogy relicts might also be expected along the Gulf Coastal Plain. The problem of detecting them is complicated by the great variety of vegetation types, the rich native fauna, and the paucity of collections from such critical areas as southern Veracruz and Tabasco. The stepwise depletion of tropical faunas between 18° and 28° N. lat. described earlier is the outstanding feature of the tropical lowlands. Nevertheless, at least a few Pleistocene relicts are evident. In eastern Mexico *Triturus (=Diemictylus) kallerti*, *Ophisaurus*, and the *oligosoma* group of *Leiolopisma* and *Terrapene mexicana* are isolated from related species in Texas. The distribution of *Coluber* south of the Río Grande suggests post-Wisconsin fragmentation and isolation of a once widespread species (see p. 67).

A disjunct lowland pattern involving dry-forest animals in the northern part of the Yucatán Peninsula and in northeastern Mexico is represented by the following: *Hypopachus cuneus*, *Laemanctus serratus*, *Sceloporus serrifer*, *Agkistrodon bilineatus*, and *Terrapene mexicana.* Unfortunately, the necessary evidence that these species do not occur at the present time in the savannas or other semiarid habitat of southern Veracruz and Tabasco is not firmly established. Perhaps the outstanding Yucatán endemic of derivation from the north is *Opheodrys mayae.* The genus *Opheodrys* occurs in both Asia and eastern North America. *O. mayae* is reported from several localities in the northern part of the Yucatán. The rough green snake, *O. aestivus*, ranges south at least to the vicinity of Tampico. Thus it is surprising to find that *O. mayae* is a relative of the geographically more remote smooth green snake, *O. vernalis*, the latter only doubtfully recorded from eastern Texas.

The apparent relationship between dry forest and scrub faunas in the Yucatán Peninsula and those in northeastern Mexico is not a simple one as the following points illustrate: (1) Unlike those of northeastern Mexico the arid Yucatán forests are an important center of endemism at the subspecific, specific, and possibly generic level. *Triprion*, a hylid frog, is among the latter. Paynter (1955) discussed the endemic element in the avifauna. (2) A tendency for animals generally regarded as humid forest dwellers to occupy dry forest or even thorn scrub is evident in the Yucatán Peninsula. Examples include: *Dactylortyx thoracicus*, *Onochorhynchus coronatus*, *Pteroglossus torquatus* (birds), *Mazama sartorii*, *Ototylomys phyllotis*, *Cuniculus paca* (mammals) and *Pliocercus elapoides*, *Ninia sebae*, and *Corythophanes cristatus* (reptiles). None of these is reported from dry forest in northeastern Mexico. (3) The majority of species shared by northeastern Mexico and northern Yucatán appear to be continuous in distribution. (4) The Arid Interior component (see p. 86), a dry continental assemblage penetrating eastern Mexico from southwestern United States, is totally unrepresented in the Yucatán Peninsula.

The history of dry forest and scrub reflects at best filtering faunal contact between northern Yucatán and northeastern Mexico. In most details the faunas of these areas are not so similar as one might expect in view of the resemblance in vegetation. Paynter (1955:318) proposed that dry-region birds may have reached the Peninsula via an arid-forest corridor extending along the eastern slope of Middle America during the last interglacial period. The concept that interglacials were hot and dry at this low latitude, however, may be less plausible than the assumption that they were moist. A dry corridor is possible during times of glacial advance when the subtropical high pressure cell over northwestern Mexico was displaced southward and when fall in sea level may have reduced significantly the evaporating surface of the Gulf of Mexico. Whatever the climatic nature of such a connection, the faunal relationships between rather similar dry forests in the Yucatán and in northeastern Mexico are more remote than we might expect.

GENERAL SUMMARY

The climatic gradient along the tropical lowlands and adjacent slopes of northeastern Mexico from 18° to 28° N. lat. involves the following four changes: (1) slight decrease in mean annual temperature, (2) appreciable increase in seasonal temperature range, (3) striking decrease in annual precipitation, and (4) a shift from a seasonal to a nonseasonal distribution of rainfall. Vegetation following this gradient is analogous to the Seasonal Formation-Series of Beard (1955), and extends from Rainforest in southern Veracruz to Thorn Scrub in southern Texas. In response to limiting climatic conditions, notably diminished rainfall, and the development of marked thermal seasons, the lowland tropical fauna undergoes depletion northward.

One section of this gradient, the Gómez Farías region in southwestern Tamaulipas, immediately south of 23° 30′ N. lat., was studied in detail. The biogeographic significance of this region stems from the fact that it is the approximate northern limit of three important plant formations, Cloud Forest, Tropical Evergreen Forest, and Tropical Deciduous Forest. The latter is spatially largest and the most influential in determining the nature of the fauna. The great number of tropical terrestrial vertebrates at or very near their northern limit in southern Tamaulipas can be related to these three tropical plant formations, especially the Tropical Deciduous Forest.

Other formations that reach their limits either north or south of the Gómez Farías region, as Thorn Scrub in southern Texas and Rainforest in southern Veracruz, may be of equal importance in preventing the northward spread of other elements of the tropical fauna. Thus the designation of any single area in the Gulf coastal gradient as the northern limit of a

"Neotropical Region" is entirely arbitrary from a faunal and ecological viewpoint. Although the Tamaulipan lowlands in the vicinity of 23° 30′ N. lat., including the Gómez Farías region, have had this distinction since the time of Wallace, this area is only one of several in which many elements of the rich Middle American tropical biota terminate.

The Gómez Farías region also lies near the southern limit of an interior continental temperate fauna. In arid interior valleys and rainshadow slopes several species penetrate the Plateau south almost to Mexico City, but in the Gulf lowlands most do not occur south of Thorn Forest in southern Tamaulipas and northern Veracruz.

Topographically, climatically, and vegetationally the Gómez Farías region is quite varied. Within the space of a few kilometers, the Sierra Madre Oriental rises from the coastal plain to 2400 m., producing a variety of climatic conditions. In response to this climatic diversity eight major plant formations are represented. A reptile and amphibian fauna of 94 species in this area can be divided on the basis of zonal distribution into three faunal groups: (1) a humid montane group in both Cloud Forest and Pine-Oak Forest; (2) a dry woodland-Thorn Scrub group surrounding the humid montane habitats; and (3) a lowland tropical group representing species that occur in subhumid forests east of the montane element, Tropical Deciduous, Tropical Evergreen, and, occasionally, Cloud Forest.

Species comprising the humid montane group in the Gómez Farías region have closely related populations in similar habitats elsewhere in the Sierra Madre Oriental. A large number are found in Humid Pine-Oak Forest in the state of Hidalgo. The present distribution of humid montane forests in northeastern Mexico, isolated in both the Gómez Farías region and elsewhere by arid corridors and bolson basins, reflects post-pluvial confinement of a habitat formerly widespread. Rainfall necessary to achieve a continuity of the now isolated montane forests is estimated at twice the present amount.

There is less convincing, but suggestive, faunal evidence for Pleistocene spread of oak savannas across the northern part of the Plateau, and of dry lowland tropical forests extending between northeastern Mexico and the northern Yucatán Peninsula.

LITERATURE CITED

Allee, Ward C.
1926 Measurement of Environmental Factors in the Tropical Rain-Forest of Panama. Ecology, 7:273-302.

Arellano, A. R. V.
1951 Research on the Continental Neogene of Mexico. Am. Journ. Sci., 249:604-616.

Baker, Rollin H.
1956 Mammals of Coahuila, Mexico. Univ. Kansas Publ. Mus. Nat. Hist., 9:127-335.

Beard, John S.
1955 The Classification of Tropical American Vegetation-Types. Ecology, 36: 89-100.

Bravo Hollis, Helia
1952 Iconographia de los cactaceas Mexicanas (Segunda Serie). Ann. Inst. Biol. Mexico, 23:501-557.

Brown, William H.
1919 Vegetation of the Philippine Mountains: The Relation Between the Environment and Physical Types at Different Altitudes. Manilla Bur. Sci. 434 pp.

Burger, W. Leslie, and William B. Robertson
1951 A New Subspecies of the Mexican Moccasin, *Agkistrodon bilineatus*. Univ. Kansas Sci. Bull., 34:213-218.

Contreras Arias, Alfonso
1942 Estudios climatologicos. Sec. Agricultura y Fomento, Direccion de Geog., Met., y Hidrologia.

Cook, O. F.
1909 Vegetation Affected by Agriculture in Central America. U. S. Dept. Agriculture Bureau of Plant Industry No. 145:1-30.

Cook, Sherburne F., and Lesley Byrd Simpson
1948 The Population of Central Mexico in the Sixteenth Century. Ibero-Americana, 31:1-241.

Crum, Howard A.
1951 The Appalachian-Ozarkian Element in the Moss Flora of Mexico with a Check-List of All Known Mexican Mosses. Microfilm Publ. #3486, Abstract, *ibid.*, V. Univ. Michigan Ph.D. thesis. 504 pp. Unpublished.

Dalquest, Walter W.
1953 Mammals of the Mexican State of San Luis Potosí. Louisiana State Univ. Studies, Biol. Sci. Ser. No. 1:1-229.

Dansereau, Pierre
1952 The Varieties of Evolutionary Opportunity. Rev. Canadienne Biol., 11:305-388.

Darling, Donald M., and Hobart M. Smith
1954 A Collection of Reptiles and Amphibians from Eastern Mexico. Trans. Kansas Acad. Sci., 57:180-195.

Darnell, Rezneat M., Jr.
1953 An Ecological Study of the Río Sabinas and Related Waters in Southern Tamaulipas, Mexico, with Special Reference to the Fishes. Univ. Minnesota Ph.D. thesis. 194 pp. Unpublished.

Davis, L. Irby
1952 Winter Bird Census at Xilitla, San Luis Potosí, Mexico. Condor, 54:345-355.

Davis, William B.
1953 Northernmost Record of the Frog *Rhinophrynus dorsalis* in Mexico. Copeia, 1953 (1):65.

Dice, Lee R.
1937 Mammals of the San Carlos Mountains and Vicinity. *In* The Geology and Biology of the San Carlos Mountains, Tamaulipas, Mexico. Univ. Mich. Studies, Sci. Ser., 12:245-268.

Dowling, Herndon G.
1951 A Taxonomic Study of the Ratsnakes, Genus *Elaphe* Fitzinger. I. The Status of the Name *Scotophis laetus* Baird and Girard (1853). Copeia, 1951 (1):39-44.
1952a A Taxonomic Study of the Ratsnakes, Genus *Elaphe* Fitzinger. II. The Subspecies of *Elaphe flavirufa* (Cope). Occ. Papers Mus. Zool. Univ. Mich., 540: 1-14.
1952b A Taxonomic Study of the Ratsnakes, Genus *Elaphe* Fitzinger. IV. A Check List of the American Forms. *Ibid.*, 541:1-12.

Duellman, William E.
1956 Frogs of the Hylid Genus *Phrynohyas* Fitzinger 1843. Misc. Publ. Mus. Zool. Univ. Mich., No. 96:1-47.

Edwards, John D.
1955 Studies of Some Early Tertiary Red Conglomerates of Central Mexico. Geol. Surv. Prof. Paper 264-H.

Hairston, Nelson G.
1949 The Local Distribution and Ecology of the Plethodontid Salamanders of the Southern Appalachians. Ecological Monog., 19:47-73.

Harrell, Byron E.
1951 The Birds of Rancho del Cielo, an Ecological Investigation in the Oak-Sweet Gum Forests of Tamaulipas, Mexico. Univ. Minnesota M. A. thesis, 283 pp. Unpublished.

Heim, Arnold
1940 The Front Ranges of the Sierra Madre Oriental, Mexico, from Ciudad Victoria to Tamazunchale. Eclogae Geologicae Helvetiae, 33:313-352.

Hernández X., Efraim, Howard Crum, William B. Fox, and Aaron J. Sharp
1951 A Unique Vegetational Area in Tamaulipas. Bull. Torrey Bot. Club, 78: 458-463.

Hill, Lawrence Francis
1926 José de Escandón and the Founding of Nuevo Santander. Ohio State Univ. Studies, Contrib. in History and Political Sci., 9:1-149.

Hooper, Emmet T.
1952 A Systematic Review of the Harvest Mice (Genus *Reithrodontomys*) of Latin America. Misc. Publ. Mus. Zool. Univ. Mich., No. 77:1-255.
1953 Notes on Mammals of Tamaulipas, Mexico. Occ. Papers Mus. Zool. Univ. Mich., 544:1-12.

Howell, Arthur H.
1929 Revision of the American Chipmunks. North American Fauna, 52:1-157.

Imlay, Ralph W.
1936 Evolution of the Coahuila Peninsula, Mexico. Part IV. Geology of the Western Part of the Sierra de Parras. Bull. Geol. Soc. America, 47:1091-1152.

Kellum, Lewis B.
1930 Similarity of Surface Geology in Front Range of Sierra Madre Oriental to Subsurface in Mexican South Fields. Bull. Amer. Assn. Petroleum Geol., 14: 73-91.
1937 Geology of the Sedimentary Rocks of the San Carlos Mountains. *In* The Geology and Biology of the San Carlos Mountains, Tamaulipas, Mexico. Univ. Mich. Studies, Sci. Ser., 12:1-98.

Klauber, Laurence M.
1952 Taxonomic Studies of the Rattlesnakes of Mainland Mexico. Bull. Zool. Soc. San Diego, 26:1-143.

Leavenworth, William C.
1946 A Preliminary Study of the Vegetation of the Region between Cerro Tancitaro and the Río Tepalcatepec, Michoacán, Mexico. Amer. Midland Nat., 36: 137-206.

Leopold, A. Starker
1950 Vegetation Zones of Mexico. Ecology, 31:507-518.

Lowery, George H., Jr., and Robert J. Newman
1951 Notes on the Ornithology of Southeastern San Luis Potosí. Wilson Bull., 63: 315-322.

Lyon, George F.
1828 A Journal of a Residence and Tour in the Republic of Mexico in the Year 1826. John Murray, London. Vol. 1, 323 pp.

Martin, Paul S.
1952 A New Subspecies of the Iguanid Lizard *Sceloporus serrifer* from Tamaulipas, Mexico. Occ. Papers Mus. Zool. Univ. Mich., 543:1-7.
1955a Herpetological Records from the Gómez Farías Region of Southwestern Tamaulipas, Mexico. Copeia, 1955, (3):173-180.
1955b Zonal Distribution of Vertebrates in a Mexican Cloud Forest. Amer. Nat., 89: 347-361.

Martin, Paul S., and Byron E. Harrell
1957 The Pleistocene History of Temperate Biotas in Mexico and Eastern United States. Ecology, 38:468-480.

Martin, Paul S., C. Richard Robins, and William B. Heed
1954 Birds and Biogeography of the Sierra de Tamaulipas, an Isolated Pine-Oak Habitat. Wilson Bull., 66:38-57.

Martinez, Maximino
1948 Los Pinos Mexicanos. Ediciones Botas, segunda ed. Mexico, D.F. 361 pp.

Mayr, Ernst
1946 History of the North American Bird Fauna. Wilson Bull., 58:3-41.

Mertens, Robert
1950 Ein neuer zaunleguan (*Sceloporus*) aus Mexiko. Wochenschrift für Aquarien-u. Terrarienkunde, 44:13-15.

Miller, Albert, and Donald H. Gould
1951 The Extensive Cold Air Outbreak of January 24-31, 1951. Monthly Weather Rev., 79:20-26.

Mills, Richard H., and Blanche B. Hull
1949 Weather Summary, Mexico; for Use with Naval Air Pilots. H. O. Publ. Washington, D.C., 532:1-220.

Milstead, William W.
1953 Geographic Variation in the Garter Snake, *Thamnophis cyrtopsis.* Texas Journ. Sci., 5:348-379.

Milstead, William W., John S. Mecham, and Haskell McClintock
1950 The Amphibians and Reptiles of the Stockton Plateau in Northern Terrell County, Texas. *Ibid.*, 2:543-562.

Miranda, Faustino
1952 La vegetacion de Chiapas. Tuxtla Gutierrez, Chiapas, Mexico. Seccion Autografica Dept. de Prensa y Turismo. Vol. 1, 334 pp.

Muir, John M.
1936 Geology of the Tampico Region, Mexico. Amer. Assoc. Petrol. Geol., 280 pp.

Muller, Cornelius H.
1937 Plants as Indicators of Climate in Northeast Mexico. Amer. Midland Nat., 18: 986-1000.
1939 Relations of the Vegetation and Climatic Types in Nuevo León, Mexico. *Ibid.*, 21:687-729.
1947 Vegetation and Climate of Coahuila, Mexico. Madroño, 9:33-57.

Paynter, Raymond A., Jr.
1955 The Ornithogeography of the Yucatán Peninsula. Peabody Museum of Natural History, Yale Univ., Bull. 9:1-347.

Peters, James A.
1951 Studies on the Lizard *Holbrookia texana* (Troschel) with Descriptions of Two New Subspecies. Occ. Papers Mus. Zool. Univ. Mich., 537:1-20.

Poinsett, Joel Roberts
1825 Notes on Mexico Made in the Autumn of 1822. London, John Miller. 298 pp.

Reese, Robert W., and I. Lester Firschein
1950 Herpetological Results of the University of Illinois Field Expedition, Spring 1949, II. Amphibia. Trans. Kansas Acad. Sci., 53:44-54.

Richards, Paul W.
1952 The Tropical Rain Forest. Cambridge Univ. Press. 450 pp.

Salvin, Osbert, and F. DuCane Godman
1889 Notes on Mexican Birds. Ibis, 1:232-243.

Santa María, fr. Vicente
1929 Estado general de las fundaciones hechas por d. José de Escandon en la colonia del Nuevo Santander. Publicaciones del Archivo General de La Nacion, 14: 1-536. Mexico.

Schimper, Andreas Franz Wilhelm
1903 Plant Geography upon a Physiological Basis. Clarendon Press, Oxford. 839 pp.

Schmidt, Karl P.
1936 Guatemalan Salamanders of the Genus *Oedipus*. Zool. Ser., Field Mus. Nat. Hist., 20:135-166.

Schuchert, Charles
1935 Historical Geology of the Antillean-Caribbean Region. Wiley and Sons, Inc., New York. 810 pp.

Shannon, Frederick A., and Hobart M. Smith
1949 Herpetological Results of the University of Illinois Field Expedition, Spring 1949. I. Introduction, Testudines, Serpentes. Trans. Kansas Acad. Sci., 52: 494-509.

Sharp, Aaron J.
1953 Notes on the Flora of Mexico: World Distribution of the Woody Dicotyledonous Families and the Origin of the Modern Vegetation. Journ. Ecol., 41:374-380.
1954 Some Pteridophytes from Tamaulipas. Amer. Fern Journ., 44:72-76.

Sharp, Aaron J., Efraim Hernández X., Howard Crum y William B. Fox
1950 Nota floristica de una asociacion importante del suroeste de Tamaulipas, Mexico. Bol. Soc. Botanica de Mexico, 11:1-4.

Shreve, Forrest
1944 Rainfall of Northern Mexico. Ecology, 25:105-111.

Simpson, Lesley Byrd
1952 Exploitation of Land in Central Mexico in the Sixteenth Century. Ibero-Americana, 36:1-91.

Smith, Hobart M.
1936 The Lizards of the *torquatus* Group of the Genus *Sceloporus* Wiegmann, 1828. Univ. Kansas Sci. Bull., 24:539-693.
1939 The Mexican and Central American Lizards of the Genus *Sceloporus*. Zool. Ser., Field Mus. Nat. Hist., 26:1-397.
1941 On the Mexican Snakes of the Genus *Pliocercus*. Proc. Biol. Soc., Washington, 54:119-124.
1942 Descriptions of New Species and Subspecies of Mexican Snakes of the Genus *Rhadinaea*. *Ibid.*, 55:185-192.
1943 Summary of the Collections of Snakes and Crocodilians Made in Mexico under the Walter Rathbone Bacon Traveling Scholarship. Proc. U. S. Nat. Mus., 93: 393-504.
1944 Snakes of the Hoogstraal Expeditions to Northern Mexico. Zool. Ser., Field Mus. Nat. Hist., 29:135-152.
1949 Herpetogeny in Mexico and Guatemala. Ann. Assn. Amer. Geographers, 39: 219-238.
1951 A New Species of *Leiolopisma* (Reptilia:Sauria) from Mexico. Univ. Kansas Sci. Bull., 34:195-200.

Smith, Hobart M., and Leonard E. Laufe
1946 A Summary of Mexican Lizards of the Genus *Ameiva*. *Ibid.*, 31:7-73.

Smith, Hobart M., and Edward H. Taylor
1945 An Annotated Checklist and Key to the Snakes of Mexico. U. S. Nat. Mus. Bull., 187:1-239.
1948 An Annotated Checklist and Key to the Amphibia of Mexico. *Ibid.*, 194:1-118.
1950 An Annotated Checklist and Key to the Reptiles of Mexico Exclusive of the Snakes. *Ibid.*, 199:1-253.

Smith, Philip W., and Donald M. Darling
1952 Results of a Herpetological Collection from Eastern Central Mexico. Herpetologica, 8:81-86.

Smith, Philip W., Hobart M. Smith, and John E. Werler
1952 Notes on a Collection of Amphibians and Reptiles from Eastern Mexico. Texas Journ. Sci., 4:251-260.

Solem, Alan
1954 Notes on Mexican Molluscs. I. Durango, Coahuila and Tamaulipas with Description of Two New *Humboldtiana*. The Nautilus, 68:3-10.

Standley, Paul C.
1920-26 Trees and Shrubs of Mexico. Contrib. U. S. Nat. Herbarium, 23 (5 parts): 1-1721.

Stirton, R. A.
1954 Late Miocene Mammals from Oaxaca, Mexico. Amer. Journ. Sci., 252: 634-638.

Stuart, Lawrence C.
1948 The Amphibians and Reptiles of Alta Verapaz, Guatemala. Misc. Publ. Mus. Zool. Univ. Mich., No. 69:1-109.

Sutton, George M., and Olin S. Pettingill, Jr.
1942 Birds of the Gómez Farías Region, Southwestern Tamaulipas. Auk, 59:1-34.

Tamayo, Jorge L.
1949 Atlas Geografico General de Mexico. Tallares Graficos de la Nacion. Mexico, D. F.

Taylor, Edward H.
1940 New Species of Mexican Anura. Univ. Kansas Sci. Bull., 26:385-405.
1942a Mexican Snakes of the Genera *Adelophis* and *Storeria*. Herpetologica, 2:75-79.
1942b New Caudata and Salientia from Mexico. Kansas Sci. Bull., 28:295-323.
1949 A Preliminary Account of the Herpetology of the State of San Luis Potosí, Mexico. *Ibid.*, 33:169-215.
1950 Second Contribution to the Herpetology of San Luis Potosí. *Ibid.*, 33:441-457.
1952 Third Contribution to the Herpetology of the Mexican State of San Luis Potosí. *Ibid.*, 34:793-815.
1953 Fourth Contribution to the Herpetology of San Luis Potosí. *Ibid.*, 35:1587-1614.

Tihen, Joe A.
1948 A New *Gerrhonotus* from San Luis Potosí. Trans. Kansas Acad. Sci., 51: 302-305.

Trapido, Harold
1944 The Snakes of the Genus *Storeria*. Amer. Midland Nat., 31:1-84.

Vivo, Jorge A., y José C. Gomez
1946 Climatología de México. Inst. Panamericano de Geografía y Historia, Publ. 19. Mexico.

Walker, Charles F.
1955a Two New Lizards of the Genus *Lepidophyma* from Tamaulipas. Occ. Papers Mus. Zool. Univ. Mich., 564:1-10.
1955b A New Salamander of the Genus *Pseudoeurycea* from Tamaulipas. *Ibid.*, 567: 1-8.
1955c A New Gartersnake (*Thamnophis*) from Tamaulipas, Mexico. Copeia, 1955 (2): 110-113.

Ward, Robert De C., and Charles F. Brooks
1938 The Climates of North America. *In* W. Koppen and R. Geiger, Handbuch der Klimatologie, 2 (pt. J.):J9-J79.

White, Stephen S.
1942 Flora of Hacienda Vista Hermosa, Nuevo León. Papers Mich. Acad. Sci., Arts, and Letters, 25:81-87.

Wright, Albert H., and Anna Allen Wright
1949 Handbook of Frogs and Toads. Comstock Publ. Co., Inc., Ithaca, New York. 640 pp.

Plate I

Fig. 1. Thorn savanna 25 km. E of Llera, 260 m., April 19, 1953. Tree at the left is *Piscidia communis*; on the right is *Yucca* sp. Before heavy grazing and irrigation much of the drier lowlands in the Gómez Farías region may have supported a similar low savanna growth.

Fig. 2. Tropical Deciduous Forest at Hygrothermograph Station 1, 2 km. ESE of Pano Ayuctle, 100 m., April 17, 1953. *Bromelia pinguin* is growing in the left foreground.

Plate II

Fig. 1. Tall palm forest south of Chamal, 140 m., April 23, 1953. The tallest palms may reach 27 m., but a continuous canopy usually does not exceed 18 m. in height.

Fig. 2. Tropical Deciduous Forest near Pano Ayuctle with a cornfield in the foreground, July, 1951. The trees still show effect of frost damage in early February of that year; new leaves appear first on central branches.

Plate III

Fig. 1. Tropical Evergreen Forest, *Iresine tomentella* in flower, 1 km. S of Aserradero del Paraiso, 420 m., April 25, 1953.

Fig. 2. View of the Sierra de Guatemala from Pano Ayuctle, 100 m., August, 1950. The cloud banner across the mountain front down to about 900 m. and covering the Cloud Forest is typical. Sugar cane and Gallery Forest of the Río Sabinas lie in the foreground. Photograph by Charles F. Walker.

Plate IV

Upper Cloud Forest at Valle de la Gruta, 1500 m., *ca.* 3 km. WNW of Rancho del Cielo, April 10, 1953. A sawmill now occupies this site.

Plate V

Cloud Forest interior near Rancho del Cielo, 1100 m., September, 1950. The heavy growth of tank bromeliads in more open parts of the forest as illustrated was virtually wiped out by the severe freeze of February, 1951. Photograph by Charles F. Walker.

Plate VI

Fig. 1. A karst rock castle in Humid Pine-Oak Forest at 1500 m., *ca.* 5 km. NW of Rancho del Cielo, May 25, 1953. Note agaves clustered on the rock.

Fig. 2. Sawmill near La Lagunita, W of Aserradero La Gloria in Humid Pine-Oak Forest, *ca.* 2000 m., June 1, 1953.

Plate VII

Fig. 1. Montane Chaparral 10 km. S of Carabanchel, 1900 m., May 2, 1953. A few trees grow in depressions and in ravines; the rest of this area is covered with oak scrub.

Fig. 2. Thorn Forest near Jaumave, 750 m., May 7, 1953, the driest part of the Gómez Farías region.

(CONTINUED FROM INSIDE FRONT COVER)

No. 34. Mollusca of Petén and North Alta Vera Paz, Guatemala. By Calvin Goodrich and Henry van der Schalie. (1937) Pp. 50, 1 plate, 1 figure, 1 map. $0.50

No. 35. A Revision of the Lamprey Genus Ichthyomyzon. By Carl L. Hubbs and Milton B. Trautman. (1937) Pp. 109, 2 plates, 5 figures, 1 map $2.00

No. 36. A Review of the Dragonflies of the Genera Neurocordulia and Platycordulia. By Francis Byers. (1937) Pp. 36, 8 plates, 4 maps $0.50

No. 37. The Birds of Brewster County, Texas. By Josselyn Van Tyne and George Miksch Sutton. (1937) Pp. 119, colored frontis., 5 plates, 1 map $1.25

No. 38. Revision of Sciurus variegatoides, a Species of Central American Squirrel. By William P. Harris, Jr. (1937) Pp. 39, 3 plates (2 colored), 3 figures, 1 map $0.50

No. 39. Faunal Relationships and Geographic Distribution of Mammals in Sonora, Mexico. By William H. Burt. (1938) Pp. 77, 26 maps $0.75

No. 40. The Naiad Fauna of the Huron River, in Southeastern Michigan. By Henry van der Schalie. (1938) Pp. 83, 12 plates, 28 figures, 18 maps $1.00

No. 41. The Life History of Henslow's Sparrow, Passerherbulus henslowi (Audubon). By A. Sidney Hyde. (1939) Pp. 72, 4 plates, 3 figures, 1 map. $0.75

No. 42. Studies of the Fishes of the Order Cyprinodontes. XVI. A Revision of the Goodeidae. By Carl L. Hubbs and C. L. Turner. (1939) Pp. 85, 5 plates $0.90

No. 43. Aquatic Mollusks of the Upper Peninsula of Michigan. By Calvin Goodrich and Henry van der Schalie. (1939) Pp. 45, 2 maps. $0.50

No. 44. The Birds of Buckeye Lake, Ohio. By Milton B. Trautman. (1940) Pp. 466, 15 plates and a frontis., 2 maps $4.50

No. 45. Territorial Behavior and Populations of Some Small Mammals in Southern Michigan. By William H. Burt. (1940) Pp. 58, 2 plates, 8 figures, 2 maps $0.50

No. 46. A Contribution to the Ecology and Faunal Relationships of the Mammals of the Davis Mountain Region, Southwestern Texas. By W. Frank Blair. (1940) Pp. 39, 3 plates, 1 map $0.35

No. 47. A Contribution to the Herpetology of the Isthmus of Tehuantepec. IV. By Norman Hartweg and James A. Oliver. (1940) Pp. 31 . . . . $0.35

No. 48. A Revision of the Black Basses (Micropterus and Huro) with Descriptions of Four New Forms. By Carl L. Hubbs and Reeve M. Bailey. (1940) Pp. 51, 6 plates, 1 figure, 2 maps $0.75

No. 49. Studies of Neotropical Colubrinae. VIII. A Revision of the Genus Dryadophis Stuart, 1939. By L. C. Stuart. (1941) Pp. 106, 4 plates, 13 figures, 4 maps. $1.15

No. 50. A Contribution to the Knowledge of Variation in Opheodrys vernalis (Harlan), with the Description of a New Subspecies. By Arnold B. Grobman. (1941) Pp. 38, 2 figures, 1 map $0.35

No. 51. Mammals of the Lava Fields and Adjoining Areas in Valencia County, New Mexico. By Emmet T. Hooper. (1941) Pp. 47, 3 plates, 1 map. $0.50

No. 52. Type Localities of Pocket Gophers of the Genus Thomomys. By Emmet T. Hooper. (1941) Pp. 26, 1 map. $0.25

No. 53. The Crane Flies (Tipulidae) of the George Reserve, Michigan. By J. Speed Rogers. (1942) Pp. 128, 8 plates, 1 map. $1.25

No. 54. The Ecology of the Orthoptera and Dermaptera of the George Reserve, Michigan. By Irving J. Cantrall. (1943) Pp. 182, 10 plates, 2 maps. $1.50

No. 55. Birds from the Gulf Lowlands of Southern Mexico. By Pierce Brodkorb. (1943) Pp. 88, 1 map $0.75

No. 56. Taxonomic and Geographic Comments on Guatemalan Salamanders of the Genus Oedipus. By L. C. Stuart. (1943) Pp. 33, 2 plates, 1 map. $0.35

No. 57. The Amnicolidae of Michigan: Distribution, Ecology, and Taxonomy. By Elmer G. Berry. (1943) Pp. 68, 9 plates, 10 figures, 10 maps. $0.85

No. 58. A Systematic Review of the Neotropical Water Rats of the Genus Nectomys (Cricetinae). By Philip Hershkovitz. (1944) Pp. 88, 4 plates, 5 figures, 2 maps $1.15

No. 59. San Francisco Bay as a Factor Influencing Speciation in Rodents. By Emmet T. Hooper. (1944) Pp. 89, 5 plates, 18 maps. $1.25

No. 60. The Fresh-Water Triclads of Michigan. By Roman Kenk. (1944) Pp. 44, 7 plates, 5 figures $0.50

No. 61. Home Range, Homing Behavior, and Migration in Turtles. By Fred R. Cagle. (1944) Pp. 34, 2 plates, 4 figures, 1 map $0.35

No. 62. Monograph of the Family Mordellidae (Coleoptera) of North America, North of Mexico. By Emil Liljeblad. (1945) Pp. 229, 7 plates $2.00

No. 63. Phylogenetic Position of the Citharidae, a Family of Flatfishes. By Carl L. Hubbs. (1945) Pp. 38, 1 figure $0.35

No. 64. Goniobasis livescens of Michigan. By Calvin Goodrich. (1945) Pp. 36, 1 plate, 1 figure, 1 map. $0.35

No. 65. Endemic Fish Fauna of Lake Waccamaw, North Carolina. By Carl L. Hubbs and Edward C. Raney. (1946) Pp. 30, 1 plate, 2 maps $0.35

No. 66. Revision of Ceratichthys, a Genus of American Cyprinid Fishes. By Carl L. Hubbs and John D. Black. (1947) Pp. 56, 2 plates, 1 figure, 2 maps $1.00

No. 67. A Small Collection of Fishes from Rio Grande do Sul, Brazil. By A. Lourenço Gomes. (1947) Pp. 39, 3 plates, 2 figures $0.50

No. 68. The Cyprinodont Fishes of the Death Valley System of Eastern California and Southwestern Nevada. By Robert R. Miller. (1948) Pp. 155, 15 plates, 5 figures, 3 maps $2.00

Printed and bound by CPI Group (UK) Ltd, Croydon, CR0 4YY

14/04/2025

14656886-0001